MARTIN GARDNER

o

The
Unexpected
Hanging

And Other
Mathematical
Diversions

A FIRESIDE BOOK
PUBLISHED BY SIMON AND SCHUSTER

A Fireside Book
Published by Simon and Schuster
A Division of Gulf & Western Corporation
Simon & Schuster Building
Rockefeller Center
1230 Avenue of the Americas
New York, New York 10020

3   4   5   6   7   8   9   10

ISBN 0-671-20073-9
ISBN 0-671-21425-X Pbk.
Library of Congress Catalog Card Number: 68-8120
Manufactured in the United States of America

For My Niece

Dorothy Elise Weaver

CONTENTS

INTRODUCTION

PIET HEIN, whose inventions in the field of recreational mathematics have many times provided material for my *Scientific American* column, "Mathematical Games," is best known in his native Denmark as the author of an enormously popular and seemingly endless series of books of short epigrammatic poems that he calls "grooks"—charmingly written, witty, and filled with the wisdom of a man who is as much at home in science and mathematics as he is in politics and the liberal arts. In the first of his book collections of poems in English, *Grooks* (Cambridge, Mass.: M.I.T. Press, 1966), the following grook appears:

> Taking fun
>   as simply fun
> and earnestness
>   in earnest
> shows how thoroughly
>   thou none
> of the two
>   discernest.

I know of no more concise way of expressing the point of view from which the chapters of this book are written. They approach mathematics in a spirit of fun, but combined with the fun there is an earnest effort to lead the reader into areas of mathematics that are far from trivial; areas that are playing essential roles in the technological revolution that so explosively is altering history and transforming our daily lives.

This is Simon and Schuster's fifth book collection of my *Scientific American* columns (including a small book, *The Numerology of Dr. Matrix*, that appeared in 1967). As in the case of previous collections, each column has been expanded, partly with material I have encountered since the column was first written, partly with material sent to me by loyal readers. At the back of the book I have listed for each chapter, excepting only those chapters which consist of short unrelated puzzles, selected references that will provide the reader with more information on the topics covered.

Each year, as my correspondence on the column grows, it becomes increasingly difficult to reply, much as I would like to do so, to every letter. Perhaps this is a good opportunity to mention three types of letters that I cannot answer:

1. I have neither the time nor competence to supply evaluations of proofs of such famous unsolved problems as the four-color-map theorem, Fermat's last theorem, and others, or to search out errors in trisections of the angle, squarings of the circle, and duplications of the cube.

2. I have not the time, and often not the competence, to provide high-school students with bibliographies and suggestions for building mathematical projects for science fairs, or for furnishing answers to difficult mathematical problems assigned to them by their teachers.

(3) Letters to me are opened at the *Scientific American* offices and shipped to me in weekly batches without their original envelopes. I am unable to answer a reader who fails to put his address on his letter or who puts it there in illegible handwriting.

I do not wish, however, to give the impression that I am not indebted to the thousands of devoted readers from all over the world who take the time to point out mistakes or inform me of unusual aspects of a topic that I failed to include, in many cases because I did not know about them. I try to reply to as many such letters as I can—if I replied to *all* of them I would have no time for anything else—but if I do not do so, let it not be thought that the letter was not read or that I am not grateful for receiving it.

MARTIN GARDNER

CHAPTER ONE

O

The Paradox
of the
Unexpected Hanging

"A NEW AND POWERFUL PARADOX has come to light." This is the opening sentence of a mind-twisting article by Michael Scriven that appeared in the July 1951 issue of the British philosophical journal *Mind*. Scriven, who bears the title of "professor of the logic of science" at the University of Indiana, is a man whose opinions on such matters are not to be taken lightly. That the paradox is indeed powerful has been amply confirmed by the fact that more than twenty articles about it have appeared in learned journals. The authors, many of whom are distinguished philosophers, disagree sharply in their attempts to resolve the paradox. Since no consensus has been reached, the paradox is still very much a controversial topic.

No one knows who first thought of it. According to the Harvard University logician W. V. Quine, who wrote one of the articles (and who discussed paradoxes in *Scientific American* for April 1962), the paradox was first circulated by word of mouth in the early 1940's. It often took the form of a puzzle about a man condemned to be hanged.

The man was sentenced on Saturday. "The hanging will take place at noon," said the judge to the prisoner, "on one of the seven days of next week. But you will not know which day it is until you are so informed on the morning of the day of the hanging."

The judge was known to be a man who always kept his word. The prisoner, accompanied by his lawyer, went back to his cell. As soon as the two men were alone the lawyer broke into a grin. "Don't you see?" he exclaimed. "The judge's sentence cannot possibly be carried out."

"I don't see," said the prisoner.

"Let me explain. They obviously can't hang you next Saturday. Saturday is the last day of the week. On Friday afternoon you would still be alive and you would know with absolute certainty that the hanging would be on Saturday. You would know this *before* you were told so on Saturday morning. That would violate the judge's decree."

"True," said the prisoner.

"Saturday, then is positively ruled out," continued the lawyer. "This leaves Friday as the last day they can hang you. But they can't hang you on Friday because by Thursday afternoon only two days would remain: Friday and Saturday. Since Saturday is not a possible day, the hanging would have to be on Friday. Your knowledge of that fact would violate

Figure 1
The prisoner eliminates all
possible days

the judge's decree again. So Friday is out. This leaves Thursday as the last possible day. But Thursday is out because if you're alive Wednesday afternoon, you'll know that Thursday is to be the day."

"I get it," said the prisoner, who was beginning to feel much better. "In exactly the same way I can rule out Wednesday, Tuesday and Monday. That leaves only tomorrow. But they can't hang me tomorrow because I know it today!"

In brief, the judge's decree seems to be self-refuting. There is nothing logically contradictory in the two statements that make up his decree; nevertheless, it cannot be carried out in practice. That is how the paradox appeared to Donald John O'Connor, a philosopher at the University of Exeter, who was the first to discuss the paradox in print (*Mind*, July 1948). O'Connor's version of the paradox concerned a military commander who announced that there would be a Class A blackout during the following week. He then defined a Class A blackout as one that the participants could not know would take place until after 6 P.M. on the day it was to occur.

"It is easy to see," wrote O'Connor, "that it follows from the announcement of this definition that the exercise cannot take place at all." That is to say, it cannot take place without violating the definition. Similar views were expressed by the authors of the next two articles (L. Jonathan Cohen in *Mind* for January 1950, and Peter Alexander in *Mind* for October 1950), and even by George Gamow and Marvin Stern when they later included the paradox (in a man-to-be-hanged form) in their book *Puzzle Math* (New York: Viking, 1958).

Now, if this were all there was to the paradox, one could agree with O'Connor that it is "rather frivolous." But, as Scriven was the first to point out, it is by no means frivolous, and for a reason that completely escaped the first three authors. To make this clear, let us return to the man in the cell. He is convinced, by what appears to be unimpeachable logic, that he cannot be hanged without contradicting the conditions specified in his sentence. Then on Thursday morning, to his great surprise, the hangman arrives. Clearly he did not expect him. What is more surprising, the judge's decree is now seen to be perfectly correct. The sentence can be carried out exactly as stated. "I think this flavour of logic refuted by the world makes the paradox rather fascinating," writes Scriven. "The logician goes pathetically through the motions that have always worked the spell before, but somehow the monster, Reality, has missed the point and advances still."

In order to grasp more clearly the very real and profound linguistic difficulties involved here, it would be wise to restate the paradox in two other equivalent forms. By doing this we can eliminate various irrelevant factors that are often raised and that cloud the issue, such as the possibility of the judge's changing his mind, of the prisoner's dying before the hanging can take place, and so on.

The first variation of the paradox, taken from Scriven's article, can be called the paradox of the unexpected egg.

Imagine that you have before you ten boxes labeled from 1 to 10. While your back is turned, a friend conceals an egg in one of the boxes. You turn around. "I want you to open these boxes one at a time," he tells you, "in serial order. Inside one of them I guarantee that you will find an unexpected egg. By 'unexpected' I mean that you will not be able to deduce which box it is in before you open the box and see it."

Assuming that your friend is absolutely trustworthy in all his statements, can his prediction be fulfilled? Apparently

Figure 2
The paradox of the unexpected egg

not. He obviously will not put the egg in box 10, because after you have found the first nine boxes empty you will be able to deduce with certainty that the egg is in the only remaining box. This would contradict your friend's statement. Box 10 is out. Now consider the situation that would arise if he were so foolish as to put the egg in box 9. You find the first eight boxes empty. Only 9 and 10 remain. The egg cannot be in box 10. Ergo it must be in 9. You open 9. Sure enough, there it is. Clearly it is an *expected* egg, and so your friend is again proved wrong. Box 9 is out. But now you have started on your inexorable slide into unreality. Box 8 can be ruled out by precisely the same logical argument, and similarly boxes 7, 6, 5, 4, 3, 2 and 1. Confident that all ten boxes are empty, you start to open them. What have we here in box 5? A totally unexpected egg! Your friend's prediction is fulfilled after all. Where did your reasoning go wrong?

To sharpen the paradox still more, we can consider it in a third form, one that can be called the paradox of the unexpected spade. Imagine that you are sitting at a card table opposite a friend who shows you that he holds in his hand the thirteen spades. He shuffles them, fans them with the faces toward him and deals a single card face down on the table. You are asked to name slowly the thirteen spades, starting with the ace and ending with the king. Each time you fail to name the card on the table he will say "No." When you name the card correctly, he will say "Yes."

"I'll wager a thousand dollars against a dime," he says, "that you will not be able to deduce the name of this card before I respond with 'Yes.'"

Assuming that your friend will do his best not to lose his money, is it possible that he placed the king of spades on the table? Obviously not. After you have named the first twelve spades, only the king will remain. You will be able to deduce the card's identity with complete confidence. Can it be the queen? No, because after you have named the jack only the king and queen remain. It cannot be the king, so it must be the queen. Again, your correct deduction would win you $1,000. The same reasoning rules out all the remaining cards. Regardless of what card it is, you should be able to deduce its name in advance. The logic seems airtight. Yet it is equally

Figure 3
The paradox of the unexpected spade

obvious, as you stare at the back of the card, that you have not the foggiest notion which spade it is!

Even if the paradox is simplified by reducing it to two days, two boxes, two cards, something highly peculiar continues to trouble the situation. Suppose your friend holds only the ace and deuce of spades. It is true that you will be able to collect your bet if the card is the deuce. Once you have named the ace and it has been eliminated you will be able to say: "I deduce that it's the deuce." This deduction rests, of course, on the truth of the statement "The card before me is either the ace or the deuce of spaces." (It is assumed by everybody, in all three paradoxes, that the man *will* be hanged, that there *is* an egg in a box, that the cards *are* the cards designated.) This is as strong a deduction as mortal man can ever make about a fact of nature. You have, therefore, the strongest possible claim to the $1,000.

Suppose, however, your friend puts down the ace of spades. Cannot you deduce at the outset that the card is the ace? Surely he would not risk his $1,000 by putting down the deuce.

Therefore it *must* be the ace. You state your conviction that it is. He says "Yes." Can you legitimately claim to have won the bet?

Curiously, you cannot, and here we touch on the heart of the mystery. Your previous deduction rested only on the premise that the card was either the ace or the deuce. The card is not the ace; therefore it is the deuce. But now your deduction rests on the same premise as before plus an additional one, namely on the assumption that your friend spoke truly; to say the same thing in pragmatic terms, on the assumption that he will do all he can to avoid paying you $1,000. But if it is possible for you to deduce that the card is the ace, he will lose his money just as surely as if he put down the deuce. Since he loses it either way, he has no rational basis for picking one card rather than the other. Once you realize this, your deduction that the card is the ace takes on an extremely shaky character. It is true that you would be wise to bet that it is the ace, because it probably is, but to win the bet you have to do more than that: you have to prove that you have deduced the card with iron logic. This you cannot do.

You are, in fact, caught up in a vicious circle of contradictions. First you assume that his prediction will be fulfilled. On this basis you deduce that the card on the table is the ace. But if it is the ace, his prediction is falsified. If his prediction cannot be trusted, you are left without a rational basis for deducing the name of the card. And if you cannot deduce the name of the card, his prediction will certainly be confirmed. Now you are right back where you started. The whole circle begins again. In this respect the situation is analogous to the vicious circularity involved in a famous card paradox first proposed by the English mathematician P. E. B. Jourdain in 1913 (*see Figure 4*). Since this sort of reasoning gets you no further than a dog gets in chasing its tail, you have no logical way of determining the name of the card on the table. Of course, you may *guess* correctly. Knowing your friend, you may decide that it is highly probable he put down the ace. But no self-respecting logician would agree that you have "deduced" the card with anything close to the logical certitude involved when you deduced that it was the deuce.

The flimsiness of your reasoning is perhaps seen more

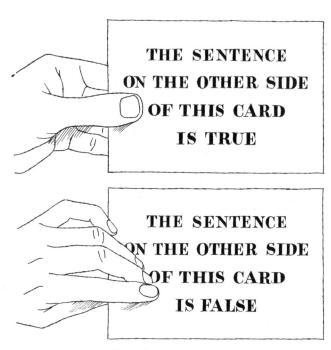

Figure 4
P.E.B. Jourdain's card paradox

clearly if you return to the ten boxes. At the start you "deduce" that the egg is in box 1, but box 1 is empty. You then
"deduce" it to be in box 2, but box 2 is empty also. Then you
"deduce" box 3, and so on. (It is almost as if the egg, just
before you look into each box in which you are positive it
must be, were cleverly transported by secret trap doors to a
box with a higher number!) Finally you find the "expected"
egg in box 8. Can you maintain that the egg is truly "expected" in the sense that your deduction is above reproach? Obviously you cannot, because your seven previous "deductions"
were based on exactly the same line of reasoning, and each
proved to be false. The plain fact is that the egg can be in any
box, *including the last one.*

Even after having opened nine empty boxes, the question
of whether you can "deduce" that there is an egg in the last
box has no unambiguous answer. If you accept only the premise that one of the boxes contains an egg, then of course an
egg in box 10 can be deduced. In that case, it is an expected
egg and the assertion that it would not be is proved false. If
you also assume that your friend spoke truly when he said

the egg would be unexpected, then nothing can be deduced, for the first premise leads to an expected egg in box 10 and the second to an unexpected egg. Since nothing can be deduced, an egg in box 10 will be unexpected and both premises will be vindicated, but this vindication cannot come until the last box is opened and an egg is found there.

We can sum up this resolution of the paradox, in its hanging form, as follows. The judge speaks truly and the condemned man reasons falsely. The very first step in his chain of reasoning—that he cannot be hanged on the last day—is faulty. Even on the evening of the next-to-last day, as explained in the previous paragraph with reference to the egg in the last box—he has no basis for a deduction. This is the main point of Quine's 1953 paper. In Quine's closing words, the condemned man should reason: "We must distinguish four cases: first, that I shall be hanged tomorrow noon and I know it now (but I do not); second, that I shall be unhanged tomorrow noon and know it now (but I do not); third, that I shall be unhanged tomorrow noon and do not know it now; and fourth, that I shall be hanged tomorrow noon and do not know it now. The latter two alternatives are the open possibilities, and the last of all would fulfill the decree. Rather than charging the judge with self-contradiction, therefore, let me suspend judgment and hope for the best."

The Scottish mathematician Thomas H. O'Beirne, in an article with the somewhat paradoxical title "Can the Unexpected *Never* Happen?" (*The New Scientist*, May 25, 1961), has given what seems to me an excellent analysis of this paradox. As O'Beirne makes clear, the key to resolving the paradox lies in recognizing that a statement about a future event can be known to be a true prediction by one person but not known to be true by another until after the event. It is easy to think of simple examples. Someone hands you a box and says: "Open it and you will find an egg inside." *He* knows that his prediction is sound, but *you* do not know it until you open the box.

The same is true in the paradox. The judge, the man who puts the egg in the box, the friend with the thirteen spades—each knows that his prediction is sound. But the prediction cannot be used to support a chain of arguments that results eventually in discrediting the prediction itself. It is this roundabout self-reference that, like the sentence on the face of Jourdain's card, tosses the monkey wrench into all attempts to prove the prediction unsound.

We can reduce the paradox to its essence by taking a cue from Scriven. Suppose a man says to his wife: "My dear, I'm going to surprise you on your birthday tomorrow by giving you a completely unexpected gift. You have no way of guessing what it is. It is that gold bracelet you saw last week in Tiffany's window."

What is the poor wife to make of this? She knows her husband to be truthful. He always keeps his promises. But if he does give her the gold bracelet, it will not be a surprise. This would falsify his prediction. And if his prediction is unsound, what *can* she deduce? Perhaps he will keep his word about giving her the bracelet but violate his word that the gift will be unexpected. On the other hand, he may keep his word about the surprise but violate it about the bracelet and give her instead, say, a new vacuum cleaner. Because of the self-refuting character of her husband's statement, she has no rational basis for choosing between these alternatives; therefore she has no rational basis for expecting the gold bracelet. It is easy to guess what happens. On her birthday she is surprised to receive a logically unexpected bracelet.

He knew all along that he could and would keep his word. *She* could not know this until after the event. A statement that yesterday appeared to be nonsense, that plunged her into an endless whirlpool of logical contradictions, has today suddenly been made perfectly true and noncontradictory by the appearance of the gold bracelet. Here in the starkest possible form is the queer verbal magic that gives to all the paradoxes we have discussed their bewildering, head-splitting charm.

ADDENDUM

A great many trenchant and sometimes bewildering letters were received from readers offering their views on how the paradox of the unexpected hanging could be resolved. Several went on to expand their views in articles that are listed in the bibliography for this chapter. (Ordinarily I give only a few select references for each chapter, but in this case it seemed that many readers would welcome as complete a listing as possible.)

Lennart Ekbom, who teaches mathematics at Östermalms College, in Stockholm, pinned down what may be the origin of the paradox. In 1943 or 1944, he wrote, the Swedish Broadcasting Company announced that a civil-defense exercise would be held the following week, and to test the efficiency of civil-defense units, no one would be able to predict, even on the morning of the day of the exercise, when it would take place. Ekbom realized that this involved a logical paradox, which he discussed with some students of mathematics and philosophy at Stockholm University. In 1947 one of these students visited Princeton, where he heard Kurt Gödel, the famous mathematician, mention a variant of the paradox. Ekbom adds that he originally believed the paradox to be older than the Swedish civil-defense announcement, but in view of Quine's statement that he first heard of the paradox in the early forties, perhaps this was its origin.

The following two letters do not attempt to explain the paradox, but offer amusing (and confusing) sidelights. Both were printed in *Scientific American*'s letters department, May 1963.

> SIRS:
> In Martin Gardner's article about the paradox of the unexpected egg he seems to have logically proved the impossibility of the egg being in any of the boxes, only to be amazed by the appearance of the egg in box 5. At first glance this truly is amazing, but on thorough analysis it can be proved that the egg will always be in box 5.
> The proof is as follows:
> Let S be the set of all statements.
> Let T be the set of all true statements.
> Every element of S (every statement) is either in the set T or in the set $C = S - T$, which is the complement of T, and not in both.
> Consider:

> (1) Every statement within this rectangle is an element of C.
> (2) The egg will always be in box 5.

Statement (1) is either in T or in C and not in both.

If (1) is in T, then it is true. But if (1) is true, it asserts correctly that every statement in the rectangle, including (1), is in C. Thus, the assumption that (1) is in T implies that (1) is in C.

Contradiction

If (1) is in C, we must consider two cases: the case that statement (2) is in C and the case that (2) is in T.

If (2) is in C, then both (1) and (2), that is, every statement in the rectangle, is an element of C. This is exactly what (1) asserts, and so (1) is true and is in T. Thus the assumption that both (1) and (2) are in C implies that (1) is in T.

Contradiction

If (2) is in T (and (1) is in C), then the assertion of (1) that every statement in the rectangle is in C is denied by the fact that (2) is in T. Therefore (1) is not true and is in C, which is entirely consistent.

The only consistent case is that in which statement (1) is in C and statement (2) is in T. Statement (2) must be true.

Therefore the egg will always be in box 5.

So you see that the discovery of the egg in box 5 is not so surprising after all.

GEORGE VARIAN
DAVID S. BIRKES

Stanford University
Stanford, Calif.

SIRS:
Martin Gardner's paradox of the man condemned to be hanged was read with extreme interest. I could not

resist noting that had our prisoner been a faithful statistician he would have preferred hanging on Wednesday, the fourth day. For if the judge had picked at random one day out of seven, then the probability that the prisoner would be required to wait x days in order to receive exactly one hanging is $p(x) = 1/7$. That is, any number of waiting days between one and seven is equally probable. This observation is a simple case of the more general hypergeometric waiting-time distribution

$$p(x) = \cdot \frac{\left[\dfrac{(x-1)!}{(x-k)!(k-1)!} \right] \cdot \left[\dfrac{(N-x)!}{(N-x-h+k)!(h-k)!} \right]}{\dfrac{N!}{(N-h)!(h!)}}$$

where $p(x)$ is the probability that x independent trials must be performed in order to obtain k successes if there are h favorable events mixed randomly among N. In our case we have $N = 7$ and (assuming one hanging is more than adequate) $h = k = 1$. Thus the "expected," or mean, value of x is $1/7 \ (1 + 2 + \ldots + 7) = 4$ days. However, I suppose we must always allow for that particularly tenacious reader who will rule out Wednesday on the grounds that it is "expected."

MILTON R. SEILER

Worthington, Ohio

CHAPTER TWO

O

Knots and
Borromean Rings

THREE CURIOUSLY INTERLOCKED RINGS, familiar to many people in this country as the trade-mark of a popular brand of beer, are shown in Figure 5. Because they appear in the coat of arms of the famous Italian Renaissance family of Borromeo they are sometimes called Borromean rings. Although the three rings cannot be separated, no two of them are linked. It is easy to see that if any one ring is taken from the set, the remaining two are not linked.

In a chapter on paper models of topological surfaces, which appears in the first *Scientific American Book of Mathematical Puzzles & Diversions,* I mentioned that I knew of no paper model of a single surface, free of self-intersection, that has three edges linked in the manner of the Borromean rings. "Perhaps," I wrote, "a clever reader can succeed in constructing one."

This challenge was first met in the fall of 1959 by David A. Huffman, associate professor of electrical engineering at the Massachusetts Institute of Technology. Huffman not only succeeded in making models of several different types of surface with Borromean edges; in doing so he also hit upon some beautifully simple methods by which one can construct

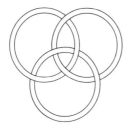

Figure 5
The three Borromean rings

paper models of a surface with edges that correspond to any type of knot or set of knots—interlaced, interwoven or linked in any manner whatever. Later he discovered that essentially the same methods have been known to topologists since the early 1930's, but because they had been described only in German publications they had escaped the attention of everyone except the specialists.

Before applying one of these methods to the Borromean rings, let us see how the method works with a less complex structure. The simplest closed curve in space is, of course, a curve that is not knotted. Mathematicians sometimes call it a knot with zero crossings, just as they sometimes call a straight line a curve with zero curvature. Diagram 1 in Figure 6 is such a curve. The shaded area in the diagram represents a two-sided surface whose edge corresponds to the curve. It is easy to cut the surface out of a sheet of paper. The actual shape of the cutout does not matter, because we are interested only in the fact that its edge is a simple closed curve. But there is another way to color the diagram. We can color the *outside* of the curve *(diagram 2 in Figure 6)* and imagine that the diagram is on the surface of a sphere. Here the closed curve surrounds a *hole* in the sphere. The two models—the first cutout and the sphere with the hole—are topologically equivalent. When put together edge to edge, they form the closed, two-sided surface of a sphere.

Now let us try the same method on a slightly more complicated diagram *(diagram 3)* of the same space curve. Think of this curve as a piece of rope. At the crossing we indicate that

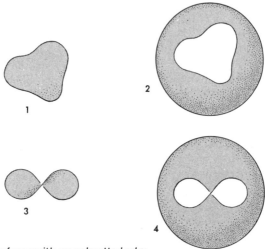

Figure 6
Models of surfaces with an unknotted edge

one segment of rope passes under the other, like a highway underpass, by breaking the line as shown. This curve also is a knot of zero crossings, because it can be manipulated so that the crossing is eliminated. (The order of a knot is the minimum number of crossings to which the knot can be reduced by deformation.) As before, we shade the diagram with two colors, tinting it so that no two regions with a common boundary have the same color. This can always be done in two different ways, one a reverse print of the other.

If we color diagram 3 as shown in the illustration, the model is merely a sheet of paper with a half twist. It is two-sided and topologically equivalent to each of the previous models. But when we color the diagram in the alternate way *(diagram 4)*, regarding the white spaces as holes in a sphere, we obtain a surface that is a Möbius strip. It too has an edge that is a knot of zero crossings (that is, not a knot), but now the surface is one-sided and topologically distinct from the preceding model. The closed, no-edged surface that results when the two models are fitted together is a cross cap, or projective plane: a one-sided surface that cannot be constructed without self-intersection.

The same general procedure can be applied to the diagram of any knot or group of knots, linked together in any manner. Let us see how it applies to the Borromean rings. The first step is to map the rings as a system of underpasses, making sure that no more than two roads cross at each pass. Next, we color the map in the two ways possible *(diagrams 1 and 2 in Figure 7).* Each crossing represents a spot where the paper surface (the shaded areas) is given a half twist in the direction indicated. The one-sided surface shown in diagram 1 is easily made with paper, either in the elegant symmetrical form shown or in topologically equivalent forms such as the one depicted in diagram 3. The model that results from diagram 2, with the Borromean rings outlining the holes in a sphere, seems at first glance quite different from the preceding model. Actually it is topologically the same. Sometimes the two methods of coloring lead to equivalent models, sometimes not.

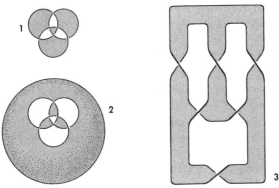

Figure 7
Topologically equivalent one-sided surfaces with Borromean-ring edges

It can be proved that this double procedure can be applied to any desired knot or group of knots, of any order, linked together in any manner. Most models obtained in this way, however, turn out to be one-sided. Sometimes it is possible to rearrange the crossings of the diagram so as to yield a two-sided surface, but usually it is extremely difficult to see how to make this sort of modification. The following method, also rediscovered by Huffman, guarantees a two-sided model.

To illustrate the procedure, let us apply it to the Borromean rings. First draw the diagram, but with light pencil lines. Place the point of the pencil on any one of the curves and trace it around, in either direction, back to the starting point. At each crossing make a small arrow to indicate the direction in which you are traveling. Do the same with each of the other two curves. The result is diagram 1 in Figure 8.

Now go over this diagram with a heavier pencil or crayon, starting at any point and moving in the direction of the arrows for that curve. Each time you come to a crossing turn either right or left as indicated by the arrows on the intersecting strand. Continue along the other strand until you reach another crossing, then turn again, and so on. It is as if you were driving on a highway and each time you reached an underpass or overpass you leaped to the other road and continued in the direction its traffic was moving. You are sure to return to your starting point after tracing out a simple closed curve. Now place the crayon at any other point on the diagram and repeat the procedure. Continue until you have gone over the entire diagram. Interestingly enough, the closed paths produced in this way will never intersect one another. In this case the result will look like diagram 2 in Figure 8.

Each closed curve represents an area of paper. Where two areas are alongside each other, the touching points represent half twists (in the direction indicated on the original diagram) that join the areas. Where one area is *inside* another, the smaller area is regarded as being above the larger, like two floor levels in a parking garage. The touching points represent half twists, but now the twists must be thought of as twisted ramps that join the two levels. The finished model is shown at 3 in Figure 8; it is two-sided and its three edges are Borromean. It can be proved that any model constructed by this procedure will be two-sided. This means that it can

Figure 8
Steps in making a two-sided surface with Borromean-ring edges

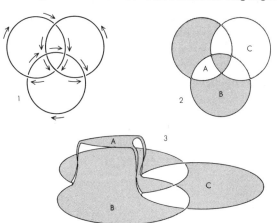

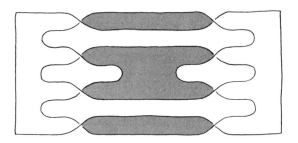

Figure 9
A two-sided, Borromean-ring-edged surface

be painted in two contrasting colors, or constructed from
paper that is differently colored on its two sides, without having one color run into the other. Figure 9, supplied by Huffman, shows a pleasingly symmetrical way of diagraming such
a surface.

The reader may enjoy building models of other knots and
linkages. The figure-of-eight knot, for example, leads to very
pleasing, symmetrical surfaces. The first diagram in Figure
10 is one way in which this familiar knot can be mapped. Diagrams of this sort, by the way, are used in knot theory for
determining the algebraic expression for a given knot. Equivalent knots, in the sense that one can be deformed into the
other, have the same algebraic formula, but not all knots with
the same formula are equivalent. It is always assumed, of
course, that the knots are tied in closed curves in three-dimensional space. Knots in ropes open at the ends, or in closed
curves in four-dimensional space, can all be untied and are
therefore equivalent to no knots at all.

The figure-of-eight knot is the only knot that reduces to a
minimum of four crossings, just as the overhand or trefoil
knot is the only type that has a minimum of three crossings.
Unlike the trefoil, however, the figure-of-eight knot has no
mirror image, or rather it can be deformed into its mirror
image. Such knots are called "amphicheiral," meaning that
they "fit either hand," like a rubber glove that can be turned
inside out.

No knots are possible with one or two crossings. There are
two five-crossers, five six-crossers, eight seven-crossers *(see*

Figure 10). This tabulation does not include mirror-image knots but does include knots that can be deformed into two simpler knots side by side. Thus the square knot *(knot 7 in the illustration)* is the "product" of a trefoil and its mirror image; the granny *(knot 8)* is the "product" of two trefoils of the same handedness. Knots 3 and 16 have very simple models. You have only to give a strip five half twists and join the ends to make its edge form knot 3, seven half twists to make it form knot 16.

All sixteen of these knots can be diagramed so that their crossings are alternately over and under. (Only knot 7, the square knot, is shown in nonalternating form.) Not until the number of crossings reach eight is it possible to construct knots (there are three) that cannot be diagramed in alternating form.

Figure 10
Knots of four crossings (1), five crossings (2, 3), six crossings (4–8) and seven (9–16)

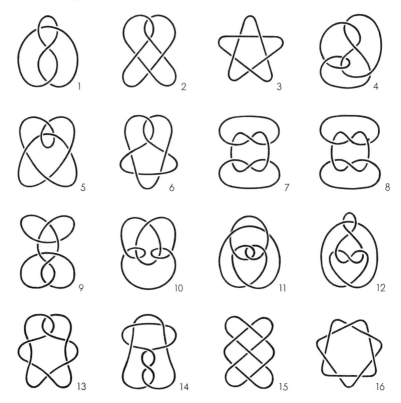

The reader may wonder why knot 9, a combination of a trefoil and a figure-of-eight, does not have two distinct forms like the square knot and granny, knots 7 and 8, each of which combines two trefoils. The answer is that the figure-of-eight part of knot 9 can be transformed to its mirror image without altering the handedness of the trefoil part, therefore there is only the knot shown and its mirror image.

A knot that cannot be deformed into simpler knots side by side is called a prime knot. All the knots in the illustration are prime except 7, 8 and 9. Knots have been carefully tabulated up through ten crossings, but no formula has yet emerged by which the number of different knots, given n crossings, can be determined. The number of prime knots with ten crossings is thought to be 167. Only wild guesses can be made as to the number of prime knots with eleven and twelve crossings.

Like topology, to which it obviously is closely related, the theory of knots is riddled with unsolved, knotty problems. There is no general method known for deciding whether or not any two given knots are equivalent, or whether they are interlocked, or even for telling whether a tangled space curve is knotted or not. To illustrate the latter difficulty, I have concocted the puzzle depicted in Figure 11. This strange-looking surface is one-sided and one-edged, like a Möbius strip, but is the edge knotted? If so, what kind of knot is it? The reader is invited to study the picture, make a guess, then test his guess by the following empirical method. Construct the surface with paper and cut it along the broken line. This will produce one single strip that will be tied in the same type of knot as the edge of the original surface. By manipulating the strip carefully so as not to tear the paper you can reduce it to its simplest form and see if your guess is verified. The result may surprise you.

In the 1860's the British physicist William Thomson (later Lord Kelvin) developed a theory in which atoms are vortex rings in an incompressible, frictionless, all-pervading ether. J. J. Thomson, another British physicist, later suggested that molecules might be the result of various knots and linkages of Lord Kelvin's vortex rings. This led to a flurry of interest in topology on the part of physicists (notably the Scottish

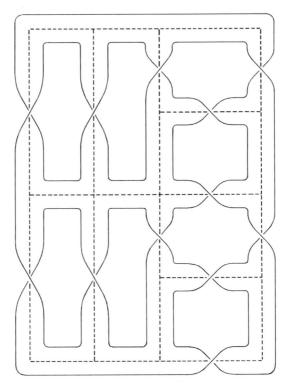

Figure 11
A one-sided, one-edged surface. Is the edge knotted?

physicist Peter Guthrie Tait), but when the vortex theory was discarded, the interest waned. Perhaps it will revive now that chemists at the Bell Telephone Laboratories have produced radically new compounds, called catenanes, that consist of carbon molecules in the form of rings that are actually linked. It is now theoretically possible to synthesize compounds made up of closed chains that can be knotted and interlocked in bizarre ways. (See Edel Wasserman, "Chemical Topology," *Scientific American,* November 1962, pages 94–102.) Who can guess what outlandish properties a carbon compound might have if all its molecules were, say, figure-of-eight knots? Or if its molecules were joined into triplets, each triplet interlocked like a set of Borromean rings?

One might suppose that living organisms would be free of knots, but such is not the case. Thomas D. Brock, a microbiologist at Indiana University, reported in *Science,* Vol. 144, No. 1620 (May 15, 1964), pages 870–72, on his discovery of a stringlike microbe that reproduces by tying itself in a knot (the knot can be an overhand, figure-of-eight, granny, or some other simple knot) which pulls tighter and tighter until the knot fuses into a bulb and free ends of the filament break off to form new microbes. And if the reader will check David Jensen's fascinating article on "The Hagfish" *(Scientific American,* February 1966, pages 82–90), he will learn about an eel-like fish that cleans itself of slime and does other curious things by tying itself into an overhand knot.

What about humans? Do they ever tie parts of their anatomy into knots? The reader is invited to fold his arms and give the matter some thought.

ANSWERS

If the surface shown in Figure 11 is constructed with paper and cut as explained, the resulting endless strip will be free of any knot. This proves that the surface's single edge is similarly unknotted. The surface was designed so that its edge corresponds to a pseudo knot known to conjurers as the Chefalo knot. It is formed by first tying a square knot, then looping one end twice through the knot in such a way that when both ends are pulled, the knot vanishes.

○

CHAPTER THREE

The Transcendental Number *e*

> *The conduct of e*
> *Is abhorrent to me.*
> *He is (not to enlarge on his disgrace)*
> *More than a little base.*
> —A Clerihew by J. A. Lindon

RECREATIONAL ASPECTS OF PI and the golden ratio, two fundamental constants of mathematics, have been discussed in previous book collections of my *Scientific American* columns. The topic of this chapter is *e*, a third great constant. It is a constant that is less familiar to laymen than the other two, but for students of higher mathematics it is a number of much greater ubiquity and significance.

The fundamental nature of *e* can best be made clear by considering ways in which a quantity can grow. Suppose you put one dollar in a bank that pays simple interest of 4 per cent a year. Each year the bank adds four cents to your dollar. At the end of 25 years your dollar will have grown to two dollars. If, however, the bank pays compound interest, the dollar will grow faster because each interest payment is added to the capital, making the next payment a trifle larger. The more often the interest is compounded, the faster the growth. If a dollar is compounded yearly, in 25 years it will grow to $(1 + 1/25)^{25}$, or \$2.66+. If it is compounded every six months (the interest is 4 per cent a year so each payment will now be 2 per cent), it will grow in 25 years to $(1 + 1/50)^{50}$, or \$2.69+.

Banks like to stress in their promotional literature the frequency with which they compound interest. This might lead one to think that if interest were compounded often enough, say a million times a year, in 25 years a dollar might grow into a sizable fortune. Far from it. In 25 years a dollar will grow to $(1 + 1/n)^n$, where n is the number of times interest is paid. As n approaches infinity, the value of this expression approaches a limit that is a mere \$2.718..., less than three cents more than what it would be if interest were compounded semiannually. This limit of 2.718... is the number e. No matter what interest the bank pays, in the same time that it would take a dollar to double in value at simple interest the dollar will reach a value of e if the interest is compounded continuously at every instant throughout the period. If the period is very long, however, even a small interest rate can grow to Gargantuan size. A dollar invested at 4 per cent in the year 1 and compounded annually would in 1960 be worth $\$1.04^{1960}$, a number of dollars that runs to about thirty-five figures.

This type of growth is unique in the following respect: at every instant its rate is proportional to the size of the growing quantity. In other words, the rate of change at any moment is always the same fraction of the quantity's value at that moment. Like a snowball tumbling down a hill, the larger it gets, the faster it expands. This is often called organic growth, because so many organic processes exhibit it. The present growth of the world's population is one dramatic example. Thousands of other natural phenomena—in physics, chemistry, biology and the social sciences—exhibit a similar type of change.

All these processes are described by formulas based on $y = e^x$. This function is so important that it is called *the* exponential function to distinguish it from other exponential functions, such as $y = 2^x$. It is a function that is exactly the same as its own derivative, a fact alone sufficient to explain e's omnipresence in the calculus. Natural logarithms, used almost exclusively in mathematical analysis (in contrast to the 10-based logarithms of the engineer), are based on e.

If you hold two ends of a flexible chain, allowing it to hang in a loop, the loop assumes the form of a catenary curve *(see Figure 12)*. The equation for this curve, in Cartesian coordinates, contains e. The cross section of sails bellying in the wind is also a catenary, the horizontal wind having the same effect on the canvas as vertical gravity on the chain. The Gilbert, Marshall and Caroline islands are the tops of volcanic sea mounts: huge masses of basalt that rest on the floor of the sea. The average profile of the mounts is a catenary. The catenary is not a conic-section curve, although it is closely related to the parabola. If you cut a parabola out of cardboard and roll it along a straight line, its focus traces a catenary.

No one has more eloquently described the catenary's appearance in nature than the French entomologist Jean Henri Fabre. "Here we have the abracadabric number e reappearing, inscribed on a spider's thread," he writes in *The Life of the Spider*. "Let us examine, on a misty morning, the mesh-

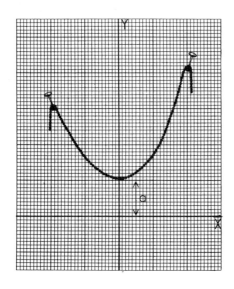

Figure 12
A chain hangs in a catenary curve. Its graph equation is:

$$y = \frac{a}{2}(e^{\frac{x}{a}} + e^{\frac{-x}{a}})$$

work that has been constructed during the night. Owing to their hygrometrical nature, the sticky threads are laden with tiny drops, and, bending under the burden, have become so many catenaries, so many chaplets arranged in exquisite order and following the curve of a swing. If the sun pierce the mist, the whole lights up with iridescent fires and becomes a resplendent cluster of diamonds. The number e is in its glory."

Like pi, e is a transcendental number: it cannot be expressed as the root of any algebraic equation with rational coefficients. Just as there is no method by which a line segment exactly equal to pi (relative to a unit segment) can be constructed with a compass and straightedge, so there is no way to construct a line segment exactly equal to e without violating the classical restraints.

Like pi, e can be expressed only as an endless continued fraction or as the limit of an infinite series. A simple way to write e as a continued fraction is:

$$e = 2 + \cfrac{1}{1 + \cfrac{1}{2 + \cfrac{2}{3 + \cfrac{3}{4 + 4}}}}$$

This continued fraction was discovered by Leonhard Euler, the eighteenth-century Swiss mathematician, who also was the first to use the symbol e. (Euler probably chose e because it was the next vowel after a, which he was using for another number, but he made so many discoveries about e that it came to be known as "Euler's number.")

If the formula $(1 + 1/n)^n$ is expanded, one obtains the following well-known infinite series that converges on e:

$$e = 1 + \frac{1}{1!} + \frac{1}{2!} + \frac{1}{3!} + \frac{1}{4!} + \frac{1}{5!} \cdots$$

The exclamation mark is the factorial sign. (Factorial 3 is $1 \times 2 \times 3$, or 6; factorial 4 is $1 \times 2 \times 3 \times 4$, or 24; and so on.) The series converges rapidly, making it as easy as pie—in fact, much easier than pi—to calculate e to any desired number of decimals. In 1952 an electronic computer at the University of Illinois, under the guiding eye of D. J. Wheeler, carried e to 60,000 decimals, and in 1961 Daniel Shanks and John W. Wrench, Jr., at the IBM Data Center in New York, extended e to 100,265 decimals! (The exclamation mark here is not a factorial sign.) Like pi, the decimals never end, nor has anyone yet detected an orderly pattern in their arrangement.

Is there a relation between e and pi, the two most famous transcendentals? Yes, many simple formulas link them together. The best known is the following formula that Euler based on an earlier discovery by Abraham de Moivre:

$$e^{i\pi} + 1 = 0$$

"Elegant, concise and full of meaning," write Edward Kasner and James R. Newman in their book *Mathematics and the Imagination*. "We can only reproduce it and not stop to inquire into its implications. It appeals equally to the mystic, the scientist, the philosopher, the mathematician." The formula unites five basic quantities: 1, 0, pi, e and i (the square root of minus one). Kasner and Newman go on to tell how this formula struck Benjamin Peirce (a Harvard mathematician and father of the philosopher Charles Sanders Peirce) with the force of a revelation. "Gentlemen," he said one day to his students after chalking the formula on the blackboard, "that is surely true, it is absolutely paradoxical; we cannot understand it, and we don't know what it means, but we have proved it, and therefore, we know it must be the truth."

Because the factorial of a number n gives the number of ways that n objects can be permuted, it is not surprising to find e popping up in probability problems that involve permutations. The classic example is the problem of the mixed-up hats. Ten men check their hats. A careless hat-check girl scrambles the checks before she hands them out. When the men later call for their hats, what is the probability of at least one man getting his own hat back? (The same problem is met in other forms. A distracted secretary puts a number of letters at random into addressed envelopes. What is the probability of at least one letter reaching the right person? All the sailors on a ship go on liberty, return inebriated and fall into bunks picked at random. What are the chances of at least one sailor sleeping in his own bunk?)

To solve this problem we must know two quantities: the number of possible permutations of 10 hats and how many of them give each man a wrong hat. The first quantity is simply 10!, or 3,628,800. But who is going to list all these permutations and then check off those that contain 10 wrong hats? Fortunately there is a simple, albeit whimsical, method of finding this number. The number of "all wrong" permutations of n objects is the integer that is the closest to $n!$ divided by e. In this case the integer is 1,334,961. The exact probability, therefore, of no man getting his hat back is 1,334,961/3,628,800, or .367879... This figure is very close to $10!/10!e$. The 10!'s cancel out, making the probability extremely close to $1/e$. This is the probability of all hats being wrong. Since it is certain that the hats are either all wrong or at least one is right, we subtract $1/e$ from 1 (*certainty*) to obtain .6321 ..., the probability of at least one man getting his own hat back. It is almost 2/3.

The odd thing about this problem is that beyond six or seven hats an increase in the number of hats has virtually no effect on the answer. The probability of one or more men getting back their hats is .6321. . . regardless of whether there are ten men or ten million men. The chart in Figure 13

Figure 13
The problem of the men and their hats

NUMBER OF HATS	NUMBER OF PERMUTATIONS	NUMBER OF PERMUTATIONS IN WHICH NO MAN GETS HAT BACK	PROBABILITY OF NO MAN GETTING HAT BACK
1	1	0	0
2	2	1	.5
3	6	2	.333333
4	24	9	.375000
5	120	44	.366666
6	720	265	.368055
7	5,040	1,854	.367857
8	40,320	14,833	.367881
9	362,880	133,496	.367879
10	3,628,800	1,334,961	.367879
11	39,916,800	14,684,570	.367879
12	479,001,600	176,214,841	.367879

shows how quickly the probability of no man getting back his hat approaches the limit of $1/e$, or .3678794411. . . The decimal fraction in the last column alternates endlessly between being a bit larger and a bit smaller.

A pleasant way to test the accuracy of all this is by playing the following game of solitaire. Shuffle a deck of cards, then deal them face up. As you deal, recite the names of all 52 cards in some previously determined order. (For example, ace to king of spades, followed by ace to king of hearts, diamonds and clubs.) You win the game if you turn up at least one card that corresponds to the card you name as you deal it. What are the chances of winning and losing?

It is easy to see that this question is identical with the question about the hats. Intuitively one feels that the probability of winning would be low—perhaps 1/2 at the most. Actually, as we have seen, it is 1 minus $1/e$, or almost 2/3. This means that in the long run you can expect to have a lucky hit about two out of every three games.

Carried to 20 decimals, e is 2.71828182845904523536. Various mnemonic sentences have been devised for remembering e, the number of letters in each word corresponding to the proper digit. In the time since I published some of these sentences (in the chapter on number memorizing in the first *Scientific American Book of Mathematical Puzzles & Diversions*) a number of readers have sent in others. Maxey Brooke of Sweeny, Texas, suggests: "I'm forming a mnemonic to remember a function in analysis." Edward Conklin of New Haven, Connecticut, went to twenty places with: "In showing a painting to probably a critical or venomous lady, anger dominates. O take guard, or she raves and shouts!" A. R. Krall, Cockeysville, Maryland, took advantage of the curious repetition of 1828 with: "He repeats: I shouldn't be tippling, I shouldn't be toppling here!"

There is a remarkable fraction 355/113 that expresses pi accurately to six decimal places. To express e to six decimals a fraction must have at least four digits above the line and four below (*e.g.*, 2721/1001). It is possible, however, to form integral fractions for e, with no more than three digits above and below the line, that give e to four decimal places. Such fractions are not so easy to come by, as the reader will quickly discover if he makes the search. For those who enjoy digital problems: What fraction with no more than three digits above the line and three below gives the best possible approximation to e?

ADDENDUM

Many readers called my attention to problems in which e turns up unexpectedly as the answer or part of the answer. I mention only two. What value of n gives the maximum value to the nth root of n? The answer is e. (See Heinrich Dörrie, *100 Great Problems of Elementary Mathematics* [New York: Dover Publications 1965], page 359.) If real numbers are picked at random from the interval 0 to 1 inclusive, and this continues until the sum of the selected numbers exceeds 1, what is the expected number of numbers that have been chosen? Again the answer is e. (See *American Mathematical Monthly*, January 1961, page 18, problem 3.)

Years ago, when I first encountered Euler's famous formula relating pi, e and the imaginary number i, I wondered if there was any way this remarkable equation could be graphed. I was unable to find a way of doing it, but L. W. H. Hull, writing on "Convergence on the Argand Diagram," in *Mathematical Gazette*, Vol. 43, No. 345 (October 1959), pages 205–7, shows how simply and elegantly it can be done. Hull first transforms the formula $e^{i\pi} = -1$ into an infinite series which is then diagramed on the complex plane as the sum of an infinite series of vectors. The i in each term of the series gives each vector a quarter turn, creating a spiral of shorter and shorter straight-line segments that strangle the point -1. A picture of the graph is reproduced in *Scientific American*, September 1964, page 59.

Concerning pi and e, a pleasant little problem that is not well known is to determine, without using tables or making actual computations, which is larger: e to the power of pi, or pi to the power of e? There are many ways to go about it, one of which is given by Phil Huneke in *The Pentagon*, Fall 1963, page 46.

ANSWERS

What integral fraction, with no more than three digits above the line and three below, gives the best possible approximation for the mathematical constant e? The answer is 878/323. In decimal form this is 2.71826. . . , the correct value for e to four decimal places. (Note to numerologists: Both numerator and denominator of the fraction are palindromes, and if the smaller is taken from the larger, the difference is 555.) Removing the last digit of each number leaves 87/32, the best approximation to e with no more than two digits in the numerator and two in the denominator.

I had hoped to be able to explain the exact technique (first called to my attention by Jack Gilbert of White Plains, New York) by which such fractions can be discovered—fractions that give the best approximations for any irrational number —but the procedure is impossible to make clear without devoting many pages to it. The interested reader will find the details in Chapter 32 of the second volume of George Chrystal's *Algebra*, a classic treatise reprinted in 1961 by Dover Publications, and in Paul D. Thomas, "Approximations to Incommensurable Numbers by Ratios of Positive Integers," *Mathematics Magazine*, Vol. 36, No. 5 (November 1963), pages 281–89.

CHAPTER FOUR

O

Geometric Dissections

MANY THOUSANDS OF YEARS AGO some primitive man surely faced, for the first time in history, a puzzling problem in geometrical dissection. Perhaps he had before him an animal skin that was large enough for a certain purpose but of the wrong shape. It had to be cut into pieces, then sewed together again in the right shape. How could it be done with the least amount of cutting and sewing? The solution of just such problems provides recreational geometry with an endlessly challenging field.

Many simple dissections were discovered by the Greeks, but the first systematic treatise on the subject seems to have been a book by Abul Wefa, a famous tenth-century Persian astronomer who lived in Baghdad. Only fragments of his book survive, but they contain gems. Figure 14 shows how Abul Wefa dissected three identical squares into nine pieces that could be reassembled to make one single square. Two squares are cut along their diagonals and the four resulting triangles are grouped around the uncut square as shown. The dotted lines show how four more cuts complete the job.

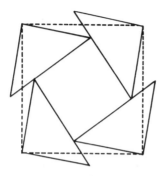

Figure 14
Nine-piece solution to Abul
Wefa's problem

It was not until this century, however, that geometers began to take seriously the task of performing such dissections in the fewest possible number of pieces. Henry Ernest Dudeney, the English puzzlist, was one of the great pioneers in this curious field. Figure 15 shows how he managed to solve Abul Wefa's three-square problem in as few as six pieces, a record that still stands.

There are several reasons why modern puzzlists have found the dissection field so fascinating. First, there is no general procedure guaranteed to work on all problems of this type,

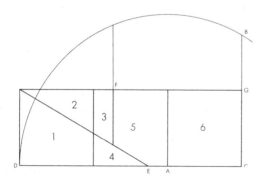

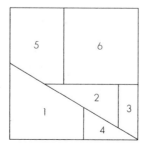

Figure 15
Six-piece solution to same
problem. Draw circle with
center at A. BC=DE=FG.

so one's intuition and creative insight are given the fullest possible play. Since no profound knowledge of geometry is called for, it is a field in which amateurs can, and in fact do, excel the professionals. Second, in most cases it has not been possible to devise a proof that a minimum dissection has actually been achieved. As a result, long-established records are constantly being shattered by new and simpler constructions.

The man who has broken more previous dissection records than anyone living today—he is the world's leading expert on such problems—is Harry Lindgren, an examiner of patents for the Australian government. He has studied all types of dissection, including plane figures with curved outlines and three-dimensional solid forms (so far as I know, no dissector has yet explored the higher dimensions!), but most of his attention has been focused on the polygons. It is not hard to prove that any polygon can be sliced into a finite number of pieces that will re-form to make any other polygon of the same area. The trick, of course, is to reduce the number of required pieces to the minimum.

The chart in Figure 16, supplied by Lindgren, shows how some of the records stood in 1961 with respect to seven of the

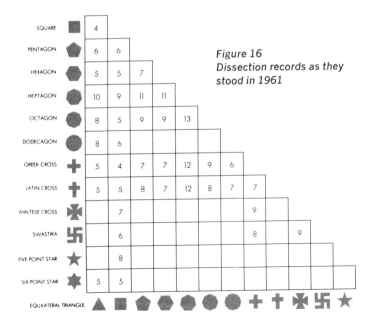

Figure 16
Dissection records as they stood in 1961

regular polygons and six other polygons of irregular but familiar shapes. The box where a row and column intersect gives the smallest number of pieces known that will form the two polygons indicated. Asymmetrical pieces can always be turned over if necessary, but a dissection is considered more elegant if this is not required. Five of the more striking of these dissections are shown in Figure 17. Four are the discoveries of Lindgren; the fifth, the dissection of a Maltese cross to a square, Dudeney attributes to one A. E. Hill. Lindgren's dissection of a hexagon to a square differs from a better-known five-piece dissection published by Dudeney in 1901. In cases such as this, where there is more than one way to obtain the minimum number of pieces, alternate dissections are almost always completely unalike. The dissection of a dodecagon to a Greek cross, published by Lindgren in *The American Mathematical Monthly* for May 1957, is one of his

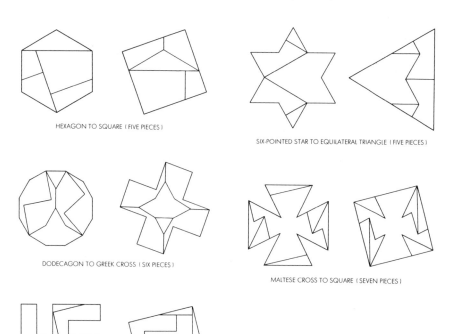

HEXAGON TO SQUARE (FIVE PIECES)

SIX-POINTED STAR TO EQUILATERAL TRIANGLE (FIVE PIECES)

DODECAGON TO GREEK CROSS (SIX PIECES)

MALTESE CROSS TO SQUARE (SEVEN PIECES)

SWASTIKA TO SQUARE (SIX PIECES)

Figure 17
Some surprising dissections

most remarkable achievements. It will be interesting to see how much this chart is altered in future years as gaps are filled in and previous records are lowered.

How does one go about trying to solve a dissection problem? It is impossible to discuss this fully here, but Lindgren has revealed his own methods in two articles ("Geometric Dissections") that appeared in *The Australian Mathematics Teacher* (Vol. 7 [1951], pages 7–10; Vol. 9 [1953], pages 17–21), and more recently in a paper entitled "Going One Better in Geometric Dissections," in the British *Mathematical Gazette* for May 1961.

One of Lindgren's methods is illustrated in Figure 18 with respect to a Latin cross and square. Each figure (the two must of course be equal in area) is first cut in some simple way so that the parts can be rearranged into a parallel-sided figure, three or four of which joined end to end form a strip with parallel sides. No cutting is necessary to form such a strip with the square (this strip is shown with broken lines), and the cross requires only one cut to form the strip drawn

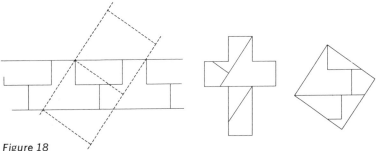

Figure 18
A five-piece dissection of a Latin cross to a square, obtained by Lindgren's strip method

with heavy lines. Both strips should be drawn on tracing paper. One is now placed on the other and turned in various ways, but always with the edges of each strip passing through what Lindgren calls "congruent points" in the pattern of the other strip. The lines that lie on the area common to both strips give a dissection of one figure into the other. The strips are tried in various positions until the best dissection is obtained. In this case the method yields the beautiful five-piece dissection shown, by which Lindgren went one better than the previous record of six.

Another method of Lindgren's can be applied if it is possible to make each polygon an element in a tessellation that fills the entire plane. By adding a small square to an octagon, for example, one obtains the tessellation shown with solid lines in Figure 19. Superposed on it is a tessellation (shown with broken lines) formed by combining a large square, its

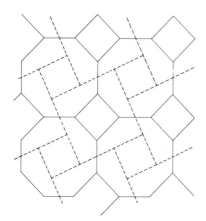

Figure 19
A five-piece dissection of an octagon to a square, obtained by Lindgren's tessellation method

area equal to that of the octagon, with a small square of the same size as before. This leads to the dissection of an octagon to a square in five pieces, a dissection first discovered by the English puzzlist James Travers and published in 1933.

Some notion of Lindgren's virtuosity is conveyed by the fact that he has managed to dissect a square into nine pieces that form either a Latin cross or an equilateral triangle; a square into nine pieces that form either a hexagon or an equilateral triangle; and a square into nine pieces that form either an octagon or a Greek cross. He has also discovered how to cut a Greek cross into twelve pieces that form three smaller Greek crosses, all alike. "Going one better in this case was not easy," he writes with understatement, referring to a previous thirteen-piece record by Dudeney. Cutting a Greek cross to form *two* smaller crosses of the same size is a much easier task that was accomplished by Dudeney with five pieces. Whether he used Lindgren's method of superposed tessellations is not known. In any case, as Lindgren points out, the Greek cross lends itself admirably to dissection by this method. By superposing two such tessellations as shown in Figure 20—one tessellation formed by repetitions of the large cross, the other by repetitions of the small one—Dudeney's solution is immediately apparent.

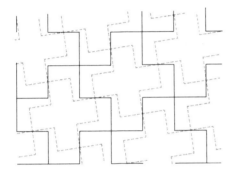

Figure 20
Tessellated Greek crosses are dissected into smaller Greek crosses

Few people can examine a dissection such as this one, Dudeney once wrote, "without being in some degree stirred by a sense of beauty. Law and order in nature are always pleasing to contemplate, but when they come under the very eye they seem to make a specially strong appeal. Even the person with no geometrical knowledge whatever is induced after the inspection of such things to exclaim, 'How very pretty!' In fact, I have known more than one person led on to a study of geometry by the fascination of cutting-out puzzles."

ADDENDUM

Since this chapter appeared in *Scientific American* in November 1961, the most significant event in the history of dissection theory has been the publication of Harry Lindgren's beautiful book *Geometric Dissections* (Princeton, N. J.: Van Nostrand, 1964). It is the only comprehensive study of dissections in any language, and likely to be the classic reference for many decades.

Figure 21 reproduces the chart shown earlier, but brought up to date (1968) by Lindgren and expanded to include 9- and 10-sided regular polygons. All these dissections may be found in Lindgren's book except for the thirteen-piece decagon to heptagon which is new. Almost all of the changes and additions to the earlier (1961) chart are due to the continued efforts of Lindgren to improve on his own earlier results. It is amusing to note that when Lindgren published a chart similar to this in his *Mathematical Gazette* article (1961), a printer's error gave six (instead of seven) as the minimum number for the Latin cross to hexagon. I corrected this on the chart that accompanied my column, but Lindgren soon confirmed the printer's figure by finding the six-piece dissection that he gives on page 20 of his book.

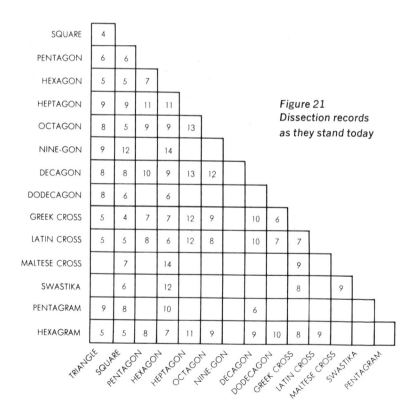

Figure 21
Dissection records
as they stand today

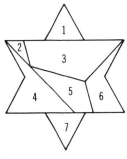
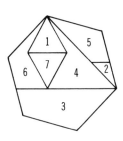

Figure 22
Bruce Gilson's dissection of six-pointed star to hexagon in seven pieces

Figure 22, a six-pointed star to hexagon in seven pieces, was first achieved in 1961 by Bruce R. Gilson, New York City, and independently by Lindgren, whose slightly different dissection is in his book on page 20. Figure 23 shows a remarkable recent discovery by Lindgren, a twelve-piece dissection of three six-pointed stars to one large star. This lowers by one the thirteen-piece dissection given in his book.

Figure 23
Lindgren's twelve-piece dissection of three six-pointed stars (each cut as shown on left) to one large star

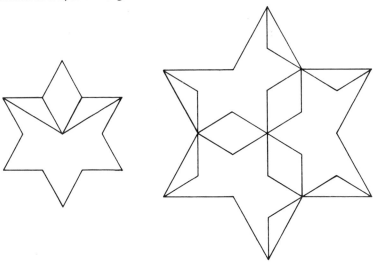

CHAPTER FIVE

O

Scarne on Gambling

From the prince's baccarat and Monte Carlo's roulette and trente-et-quarante, *to the soldier's crown and anchor and the errand boy's pitch and toss, it is a history of stakes lost, relieved by incidents of irrational acquisition. It is a history of landslides in an account book. It is a pattern of slithering cards, dancing dice, spinning roulette wheels, coloured counters and scribbled computations on a background of green baize. It is a world parasitic on the general economic organization—fungoid and aimless, rather than cancerous and destructive, in its character. A stronger, happier organization would reabsorb it or slough it off altogether.*

—H. G. WELLS,
The Work, Wealth and Happiness of Mankind

CONJURING IS THE ART OF ENTERTAINING people by performing feats that seem to violate the laws of nature. The deception is accomplished by a prodigious variety of subtle techniques, all in good fun because the ultimate intent of the performance is to delight an audience. There are, however,

two large fields of public deception in which many of the principles of magic are employed for less wholesome purposes:
the fields of gambling and psychic research. A certain false
shuffle, for instance, can be equally useful to a card magician
and to a card hustler. A technique for secretly obtaining information written on a piece of paper can be equally useful
to a magician who performs "mental magic" and to a crooked
medium. As a mathematician might put it, the principles of
deception in the three areas—magic, gambling and psychic
phenomena—form three mutually intersecting sets.

The violation of any natural law, including the mathematical laws of probability, can provide the basis for a magic
trick. One of the most famous modern card tricks, known to
magicians as Out of this World (it was invented by Paul
Curry, a New York City amateur), appears as follows: A
shuffled deck is dealt randomly, by a spectator, into two piles.
When the piles are turned over, one is seen to consist entirely of red cards, the other entirely of black. The laws of
probability have clearly been evaded and everyone is pleasant-
ly amazed. The relation between a trick of this type and deception in the psychic and gambling areas is at once apparent.
If a spectator accomplished such a feat by clairvoyance, the
feat would move into the realm of extrasensory perception
(ESP). On the other hand, if the magician achieved the results by sleight of hand, who would want to play poker with
him?

Techniques of deception in modern psychic research, which
relies almost exclusively on experiments that seem to counter
the laws of probability, have been dealt with at length by
Mark Hansel, a British psychologist, in his eye-opening book
ESP: A Scientific Evaluation (New York: Scribner's 1966).
The techniques of deception in modern gambling, again with
a major emphasis on probability laws, receive their most
comprehensive coverage in a 713-page book entitled *Scarne's
Complete Guide to Gambling*, published in 1961 by Simon and
Schuster. The book was well timed. A United States Senate
subcommittee, before which Scarne was the government's
first star witness, was at that time conducting a nationwide
investigation of illegal gambling that led to new control legis-
lation. (See *Time*, September 1, 1961, page 16.)

No one alive today is better qualified to write such a book than John Scarne. A native of Fairview, New Jersey, he developed a passionate interest in magic—card magic in particular—at an early age and quickly became one of the nation's most skillful card manipulators. The magic periodicals of twenty or thirty years ago are filled with references to Scarne's exploits and original magic creations. During the past two decades he has made an intensive study of gambling in all its multifarious and nefarious phases.

In addition to knowing all that one man can know about gambling, and having mastered the most difficult "moves" of the card and dice hustlers, Scarne has also managed to acquire a remarkable knowledge of basic probability theory. This flourishing branch of modern mathematics actually had its origin in gambling questions. Scarne retells in his book the story of how Galileo became interested in probability when an Italian nobleman asked him: When three dice are tossed, why does the total 10 show more often than the total 9? Galileo answered by making a table of the 216 equally probable ways that three dice can fall. Scarne also recounts the more familiar story of how in 1654 Blaise Pascal was asked by Antoine Chevalier de Méré (a French courtier and writer, not at all the professional gambler he is usually made out to be) why he had been losing consistently when he bet even money that a double six would show at least once in 24 rolls of two dice. Pascal was able to show that the odds were slightly against the gambler if he rolled 24 times, but would be slightly in his favor if he rolled 25 times. (For Pascal's proof, the reader is referred to Oystein Ore's essay on "Pascal and the Invention of Probability Theory" in *The Colorado College Studies*, No. 3, Spring 1959.)

Scarne tells an amusing story about how a New York City gambler called "Fat the Butch" once lost $49,000 by repeatedly betting at even odds that he could roll a double six in 21 rolls. Since there are 36 combinations with two dice, and since a double six can be made in only one way, Fat the Butch reasoned (as did many gamblers in Pascal's time) that in the long run he could expect to roll a six in 18 rolls as often as he would fail to roll a six. Since it seemed an even bet in 18 rolls, how could he lose in 21?

Scarne's book contains a complex analysis of blackjack, the only casino game in which, at certain times, the odds actually favor the player. Scarne gives the details of a sound method by which one can beat the house percentage *if* one is willing to spend time mastering the game's mathematics, becoming familiar with casino practices and, above all, learning to "case the deck" (remember every card dealt or exposed). Do you know why it is advantageous for a blackjack dealer to be left-handed? The asymmetrical position of the indices on the cards makes it easier for a left-hander to take an undetectable peek at the top card before he deals it. Another section of the book gives the full and hitherto unpublished history of the famous "rhythm method" by which slot machines in the United States were successfully bilked in the late 1940's of millions of dollars before the manufacturers understood what was happening and added a "variator" to the machines.

The book contains masterly analyses of the mathematics of bingo, poker, gin rummy, the numbers game, craps, horse racing and many other popular forms of gambling. An entire chapter is devoted to the match game, a guessing game that for many years was a favorite pastime at Bleeck's, more formally known as Artists and Writers Restaurant, in New York City. Even carnival games, including the latest swindle, a marble roll-down called Razzle Dazzle, are covered. Is there anyone who hasn't at some time thrown baseballs at those pyramids of six pint-size wooden milk bottles? The "gaff" couldn't be simpler. Three bottles are heavy; three are light. With the heavy bottles on top, one ball will topple all of them off the small table on which they stand. With the heavies on the bottom, a big-league pitcher couldn't do it. Carnival people call this a "two-way store," meaning that it can be set up for the player to win or lose.

One of the book's best chapters is on roulette, certainly the most glamorous of all casino games. "A great part of roulette's fascination," Scarne writes in a bit of purple prose, "lies in the beauty and color of the game. The surface of the handsome mahogany table is covered with a blazing green cloth which bears the bright gold, red and black of the layout. The chromium separators between the numbered pockets on the wheel's rim glitter and dance in the bright light as the

wheel spins. The varied colors of the wheel checks stacked before the croupier and scattered on the layout's betting spaces,
the evening clothes of the women, the formal dress of the
men, the courteous croupiers—all add to the enticing picture."

So enticing, indeed, that thousands of players, with limited
knowledge of probability, each year develop into what Scarne
terms "roulette degenerates." Many of them squander their
time and fortunes on worthless systems by which they expect
to break the bank and retire for life. As almost everyone
knows, a roulette wheel has on its rim the numbers from 1 to
36, plus a zero and double zero (European and South American wheels have only the zero). These numbers are not arranged randomly but are cleverly patterned to provide a maximum balance among high-low, even-odd and red-black. The
house pays off a winning number at 35 to 1, which would
make roulette an even-up game if it were not for the extra 0
and 00. These two numbers give the house a percentage on all
bets (except one) of 5 and 5/19 per cent, or about 26 cents on
every $5 bet. The one exception is the five-number line bet, in
which the chips are placed on the end of the line that separates
1, 2 and 3 from the 0 and 00 spaces on the layout board (*see
Figure 24*) ; in other words, a bet that one of these five numbers will show. Here the bank pays off at 6-to-1 odds that give
it a percentage of 7 and 17/19 per cent, or about 39 cents on
a $5 bet. Obviously this is a bad bet and one to be avoided.

The fact that every possible bet on the layout is in the
bank's favor underlies a very simple proof that no roulette
system is worth, as Scarne puts it, the price of yesterday's
newspaper. "When you make a bet at less than the correct
odds," he writes, "which you always do in any organized gambling operation, you are paying the operator a percentage
charge for the privilege of making the bet. Your chance of
winning has what mathematicians call a 'minus expectation.'
When you use a system you make a series of bets, each of
which has a minus expectation. There is no way of adding mi-
nuses to get a plus. . . ." For those who understand, this is as
ironclad and unanswerable as the impossibility proofs of the
trisection of the angle, squaring the circle and the duplication
of the cube.

The most popular of all systems, says Scarne, is known as

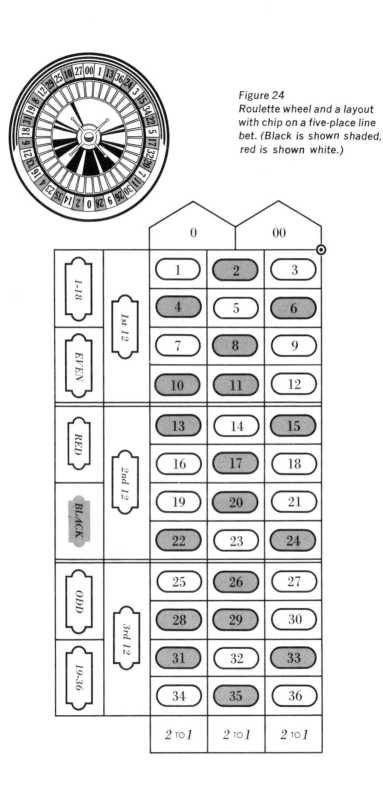

Figure 24
Roulette wheel and a layout
with chip on a five-place line
bet. (Black is shown shaded,
red is shown white.)

the d'Alembert system. It consists of betting the red or black color (or any other even-money bet), then following with a larger bet after each loss and a smaller bet after each win. The assumption is that if the little ivory ball drops several times into red, it will somehow remember this fact and tend to avoid red on the next spin. Mathematicians know this as the "gambler's fallacy," and of course it gives the player no advantage whatever.

The martingale system, in which bets are doubled until one wins, might work (in a sense) if the house did not have a limit on the size of bets. (The limit is usually the total reached by doubling the minimum required bet about seven times.) It is true that the player of the martingale has a high chance of winning a picayune amount (on even-money bets, $1 for every chain of doubling that begins with $1 and ends before the limit is reached), but this is balanced by the probability of losing a whopping amount. If you start with, say, a $180 bankroll and the bank has a maximum betting limit of $180 on the colors, the odds are greatly in your favor that you will start off a winner with the martingale. But if you continue to play it, you can expect to hit (sooner than you would expect) a seven- or eight-number losing streak that will wipe out your capital. It is as if you faced a thousand boxes, each containing a dollar bill except for one box holding a bomb that will explode when you open the box. You are allowed to open boxes at random and keep everything you find inside. The chances are very good that if you open ten boxes you will become richer by ten dollars. But is it a wise bet? Against the high probability of winning a dollar each time you must weigh the low probability of blowing yourself up. Most people would consider the situation (which has a grim analogy today with many decisions of foreign policy) one of "minus expectation."

There is a reverse version of the martingale, known in the United States as the parlay system (Europeans call it the paroli), in which you continue to bet a dollar after each previously lost bet and double your wager after each winning bet. Instead of making small wins at the risk of a big loss, you sacrifice small losses for the possibility of that one ecstatic moment when a lucky run of wins will pyramid your stake into a fortune of some previously specified amount. Here again, the system has a minus expectation even without the house limit, but the limit makes it even worse. If you started your parlay with a dollar, your doubling procedure would be permitted to go only as far as the seventh win of $128.

Another popular system, called the cancellation system, has lost fortunes for many a "mark" (sucker) who thought he had something sure-fire. Even-money bets are made (say on red or black), with the amount increased after each loss according to the following procedure. First you write down a column of figures in serial order, say from 1 to 10. Your first bet is the total of the top and bottom figures, in this case 11. If you win, cross out the 1 and 10 and bet the new top and bottom figures, 2 and 9, which also total 11. If you lose, write the loss (11) at the bottom of the row and bet the total of the new top and bottom figures, 1 and 11. This procedure continues with two numbers crossed off at each win and one number added at each loss. Since the losses are about equal to the wins on even-money bets, you are almost sure to cross off all the numbers. When this occurs, you will be 55 chips ahead!

"On paper it looks good," comments Scarne, but alas, it is merely one of the many worthless variations on the martingale. The player keeps risking larger and larger losses to win piddling amounts. But in this case the bets are smaller so it takes longer to be stopped by the house limit. Meanwhile the house's 5 and 5/19 per cent is taking its toll, and the croupier is getting increasingly annoyed at having to handle so many small bets.

One of Scarne's best gambling stories (the book is filled with them) is about an elderly drunk in a Houston casino who complained of having lost a ten-dollar "mental bet" on Number 26. He had placed no chips on the table, but since he had made the bet in his mind, he insisted on paying the croupier ten dollars before he disappeared into the bar. Back he wobbled later, watched the ball drop, then shouted excitedly: "That's me! I won!" The drunk kicked up such a commotion, demanding payment for his mind bet, that the manager had to be summoned. The manager ruled that since the croupier had accepted ten dollars on a *lost* mental bet, he had to pay off on the *winning* one. The drunk, suddenly sober, walked off with $350. "Don't try this," Scarne adds. "Everybody in the casino business knows about it."

The latest roulette system to make a big splash is one that

appeared in 1959 in *Bohemia,* a Cuban monthly magazine. For many months it was widely played throughout South America. The system is based on the fact that the third column of the layout (*see Figure 24*) has eight red numbers and only four black—a fact that the inventor of the system considers to be a fatal flaw in the layout.

Here is Scarne's description of how the system operates:

"You make two bets on each spin of the wheel. Bet one $1 chip on the color Black, which pays even money. And bet one $1 chip on the third column, which contains 8 red numbers, 3, 9, 12, 18, 21, 27, 30 and 36, and the black numbers 6, 15, 24 and 33. This bet pays odds of 2 to 1.

"There are 36 numbers plus the 0 and 00 on the layout. Suppose you make 38 two-chip bets for a total of $76. In the long run this should happen:

"1. The zero or double zero will appear 2 times in 38, and you lose 2 chips each time—a loss of 4 chips.

"2. Red will appear 18 times out of 38. Each time one of the 10 red numbers listed in the first and second column appears you lose 2 chips—a loss of 20 chips on those 10 numbers. But when the 8 red numbers in the third column appear you win 2 chips on each for a total win of 16 chips. This gives you a net loss on Red of 4 chips.

"3. Black will also appear 18 out of 38 times. Each time one of the 14 (black) numbers in the first and second column appears, you lose 1 chip—a total loss of 14 chips. But since you also bet on the color Black, you win 14 times for a gain of 14 chips. This loss and gain cancel out and you break even 14 times. But when the 4 black numbers in the third column (6, 15, 24 and 33) appear, you win 3 chips each time (2 chips on the number and 1 chip on the color) for a profit of 12 chips on Black.

"Having lost 4 chips on the zero and double zero, and 4 chips on Red, and having won 12 chips on Black, you come out ahead with a final profit of 4 chips. Divide your total bet of $76 into your profit of $4 and you will find that you have not only overcome the house advantage on the zero and double zero of 5 and 5/19 per cent but have actually supplanted it with an advantage in *your* favor of 5 and 5/19 per cent."

The reader will find it a stimulating exercise in elementary

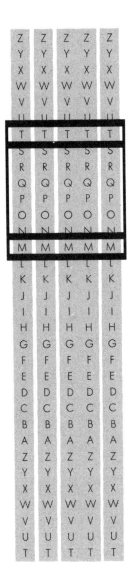

Figure 25
Find the Christmas message

probability analysis to see if he can spot the fallacy in the system.

In keeping with our occasional practice of giving a cryptic Christmas greeting, the reader is invited to study the device shown in Figure 25. The problem is to slide the four vertical strips of letters up and down until two words conveying the message appear simultaneously, one in each of the two horizontal windows. The puzzle was devised by Leigh Mercer of London.

ADDENDUM

Since Scarne published his *Complete Guide to Gambling,*
Edward O. Thorp, a mathematician at New Mexico State
University, brought out *Beat the Dealer* (New York: Blaisdell, 1962), a book in which a winning blackjack strategy is
explained in fascinating detail. The book also contains an
eye-opening account of Thorp's experiences in playing his
system in the casinos of Reno and Las Vegas. As one would
have expected, the casinos have since altered their rules to
plug some of the holes Thorp exploited. (See *The New York
Times,* April 3, 1964, first page of second section.)

Thorp's Chapter 7, on cheating, should be read by every
innocent layman who believes the widespread myth that
dealers in the major clubs of Nevada never cheat. Because
the house has a good percentage in honest play, so the myth
goes, and because cheating would only frighten away customers if it became known, the Nevada houses are the most
honest in the world. The facts are that cheating constantly
occurs in even the best casinos. The most ·common type is
cheating *against* the house on the part of a dishonest dealer
who later splits the take with a confederate. To make his
record look honest for the day, he cuts down wins to other
customers. Since the house is always suspicious of this type
of skullduggery, a dealer will sometimes prevent a large loss
to the house just so he won't be *suspected* of throwing money
to a confederate. A skillful dealer also takes great pride in
his craft and will cheat just to keep in practice and for the
fun of it. Finally, as Thorp's chapter on cheating makes
clear, many houses instruct their dealers to cheat whenever
the stakes get high. Nevada does have an inspection squad,
but it is small and inefficient. Even so, Nevada houses are
often caught cheating, although the news seldom leaks out.
The New York Times, April 12, 1964, reported the closing
of a top casino on the Las Vegas Strip after a random check
on its dice turned up a set of five "percentage dice," cubes
with edges rounded in such a way as to increase the odds
against the roller. The same casino had fired a blackjack
dealer the previous year after a state undercover agent found
him cheating. On October 17, 1967, *The New York Times*

reported the closing of a major Nevada casino after the state caught a dice dealer switching honest dice (by means of a double apron) for "mis-spot dice" that had certain numbers missing. It was the second big casino to be closed for similar reasons within a month. In 1961, when I reviewed Scarne's book in my column in *Scientific American,* two Vegas clubs were closed because of cheating.

An interesting paradox concerning the roulette layout was called to my attention by Thomas H. O'Beirne, of Glasgow, who in turn was told about it by the Polish mathematician Hugo Steinhaus. Since there are the same number of red numbers as black, of high numbers as low, and odd numbers as even, the chances of winning any single bet of this type are obviously equal. The house gets a percentage because of the 0 and 00, but otherwise the ball is just as likely to fall odd, or high, as it is to fall red. If you make a permissible pair-bet, say odd-black or even-red, your chances on each bet will be the same regardless of what pair you choose. There is no way to make a triple-bet, such as low-red-odd, but suppose such a bet could be made. Instead of paying off independently on each part of the bet, the bank pays only if the ball lands on a number that is low, red and odd. Otherwise you lose. Would your chance of winning be the same if you were to bet on, say, low-red-even? Surprisingly, it would not. This is easily seen by a careful inspection of the layout. There are five low, red, even numbers but only four low, red, odd numbers. Of the eight possible triplets, half have winning probabilities of 4/38, half have winning probabilities of 5/38.

ANSWERS

Readers were asked to spot the fallacy in a roulette system that swept the casinos of South America. Here is the answer as given in *Scarne's Complete Guide to Gambling.*

"The joker is in the statement that 'when the 8 red numbers in the third column appear, you win 2 chips on each for a total win of 16 chips.' This is incomplete. When those 8 numbers win and pay off 16 chips, *you also lose 8 chips on*

Black, making the net payoff only 8 chips. Since you lost 20 chips on the red numbers in the first and second columns, your net loss on Red is not 4 chips, as stated, but 12 chips. Having lost 12 chips on Red, won 12 chips on Black, and lost 4 chips on the zero and double zero, you end up losing 4 chips. And that washes that system out completely. The house still has its favorable edge of 5 and 5/19 per cent, as usual, and the casino operator is the guy who is going to get rich —not you."

And here is how one reader, John Stout of New York City, put it:

> *The fallacy, it must be said,*
> *Lies in those third-column wins on red,*
> *Since each of these gains but one chip,*
> *A black-win wager's lost per trip.*
> *If into this one cares to delve,*
> *He sees the net red loss as twelve.*
> *Four dollars is the* total *loss.*
> *The house still gets its five per cent,*
> *The "zero" debits make their dent,*
> *And math is still the gambler's boss.*

Those four paper strips can be adjusted to spell DOLLS WHEEL, but this fails to qualify as a Christmas greeting. The correct answer is JOLLY CHEER.

CHAPTER SIX

O

The Church
of the
Fourth Dimension

"Could I but rotate my arm out of the limits set to it," one
of the Utopians had said to him, "I could thrust it into a
thousand dimensions."

—H. G. WELLS, *Men Like Gods*

ALEXANDER POPE ONCE DESCRIBED LONDON as a "dear, droll,
distracting town." Who would disagree? Even with respect
to recreational mathematics, I have yet to make an imaginary visit to London without coming on something quite extra-
ordinary. Last fall, for instance, I was reading the London
Times in my hotel room a few blocks from Piccadilly Circus
when a small advertisement caught my eye:

WEARY OF THE WORLD OF THREE DIMENSIONS? COME WORSHIP
WITH US SUNDAY AT THE CHURCH OF THE FOURTH DIMENSION.
SERVICES PROMPTLY AT 11 A.M., IN PLATO'S GROTTO. REVEREND
ARTHUR SLADE, MINISTER.

An address was given. I tore out the advertisement, and on
the following Sunday morning rode the Underground to a

station within walking distance of the church. There was a damp chill in the air and a light mist was drifting in from the sea. I turned the last corner, completely unprepared for the strange edifice that loomed ahead of me. Four enormous cubes were stacked in one column, with four cantilevered cubes jutting in four directions from the exposed faces of the third cube from the ground. I recognized the structure at once as an unfolded hypercube. Just as the six square faces of a cube can be cut along seven lines and unfolded to make a two-dimensional Latin cross (a popular floor plan for medieval churches), so the eight cubical hyperfaces of a four-dimensional cube can be cut along seventeen squares and "unfolded" to form a three-dimensional Latin cross.

A smiling young woman standing inside the portal directed me to a stairway. It spiraled down into a basement auditorium that I can only describe as a motion-picture theater combined with a limestone cavern. The front wall was a solid expanse of white. Formations of translucent pink stalactites glowed brightly on the ceiling, flooding the grotto with a rosy light. Huge stalagmites surrounded the room at the sides and back. Electronic organ music, like the score of a science-fiction film, surged into the room from all directions. I touched one of the stalagmites. It vibrated beneath my fingers like the cold key of a stone xylophone.

The strange music continued for ten minutes or more after I had taken a seat, then slowly softened as the overhead light began to dim. At the same time I became aware of a source of bluish light at the rear of the grotto. It grew more intense, casting sharp shadows of the heads of the congregation on the lower part of the white wall ahead. I turned around and saw an almost blinding point of light that appeared to come from an enormous distance.

The music faded into silence as the grotto became completely dark except for the brilliantly illuminated front wall. The shadow of the minister rose before us. After announcing the text as Ephesians, Chapter 3, verses 17 and 18, he began to read in low, resonant tones that seemed to come directly from the shadow's head: ". . . that ye, being rooted and grounded in love, may be able to comprehend with all saints what is the breadth, and length, and depth, and height. . . ."

It was too dark for note-taking, but the following para-
graphs summarize accurately, I think, the burden of Slade's
remarkable sermon.

Our cosmos—the world we see, hear, feel—is the three-
dimensional "surface" of a vast, four-dimensional sea. The
ability to visualize, to comprehend intuitively, this "wholly
other" world of higher space is given in each century only to
a few chosen seers. For the rest of us, we must approach
hyperspace indirectly, by way of analogy. Imagine a Flatland,
a shadow world of two dimensions like the shadows on the
wall of Plato's famous cave *(Republic,* Chapter 7). But shadows do not have material substance, so it is best to think of
Flatland as possessing an infinitesimal thickness equal to the
diameter of one of its fundamental particles. Imagine these
particles floating on the smooth surface of a liquid. They
dance in obedience to two-dimensional laws. The inhabitants
of Flatland, who are made up of these particles, cannot con-
ceive of a third direction perpendicular to the two they know.

We, however, who live in three-space can see every particle of Flatland. We see inside its houses, inside the bodies
of every Flatlander. We can touch every particle of their
world without passing our finger through their space. If we
lift a Flatlander out of a locked room, it seems to him a
miracle.

In an analogous way, Slade continued, our world of three-
space floats on the quiet surface of a gigantic hyperocean;
perhaps, as Einstein once suggested, on an immense hypersphere. The four-dimensional thickness of our world is ap-
proximately the diameter of a fundamental particle. The laws
of our world are the "surface tensions" of the hypersea. The
surface of this sea is uniform, otherwise our laws would not
be uniform. A slight curvature of the sea's surface accounts
for the slight, constant curvature of our space-time. Time
exists also in hyperspace. If time is regarded as our fourth
coordinate, then the hyperworld is a world of five dimensions.
Electromagnetic waves are vibrations on the surface of the
hypersea. Only in this way, Slade emphasized, can science
escape the paradox of an empty space capable of transmit-
ting energy.

What lies outside the sea's surface? The wholly other world
of God! No longer is theology embarrassed by the contradic-

tion between God's immanence and transcendence. Hyperspace touches every point of three-space. God is closer to us than our breathing. He can see every portion of our world, touch every particle without moving a finger through our space. Yet the Kingdom of God is completely "outside" of three-space, in a direction in which we cannot even point.

The cosmos was created billions of years ago when God poured (Slade paused to say that he spoke metaphorically) on the surface of the hypersea an enormous quantity of hyperparticles with asymmetric three-dimensional cross sections. Some of these particles fell into three-space in right-handed form to become neutrons, the others in left-handed form to become antineutrons. Pairs of opposite parity annihilated each other in a great primeval explosion, but a slight preponderance of hyperparticles happened to fall as neutrons and this excess remained. Most of these neutrons split into protons and electrons to form hydrogen. So began the evolution of our "one-sided" material world. The explosion caused a spreading of particles. To maintain this expanding universe in a reasonably steady state, God renews its matter at intervals by dipping his fingers into his supply of hyperparticles and flicking them toward the sea. Those which fall as antineutrons are annihilated, those which fall as neutrons remain. Whenever an antiparticle is created in the laboratory, we witness an actual "turning over" of an asymmetric particle in the same way that one can reverse in three-space an asymmetric two-dimensional pattern of cardboard. Thus the production of antiparticles provides an empirical proof of the reality of four-space.

Slade brought his sermon to a close by reading from the recently discovered Gnostic Gospel of Thomas: "If those who lead you say to you: Behold the kingdom is in heaven, then the birds will precede you. If they say to you that it is in the sea, then the fish will precede you. But the kingdom is within you and it is outside of you."

Again the unearthly organ music. The blue light vanished, plunging the cavern into total blackness. Slowly the pink stalactites overhead began to glow, and I blinked my eyes, dazzled to find myself back in three-space.

Slade, a tall man with iron-gray hair and a small dark

mustache, was standing at the grotto's entrance to greet the members of his congregation. As we shook hands I introduced myself and mentioned this department. "Of course!" he exclaimed. "I have some of your books. Are you in a hurry? If you wait a bit, we'll have a chance to chat."

After the last handshake Slade led me to a second spiral stairway of opposite handedness from the one on which I had descended earlier. It carried us to the pastor's study in the top cube of the church. Elaborate models, three-space projections of various types of hyperstructures, were on display around the room. On one wall hung a large reproduction of Salvador Dali's painting "Corpus Hypercubus." In the picture, above a flat surface of checkered squares, floats a three-dimensional cross of eight cubes; an unfolded hypercube identical in structure with the church in which I was standing.

"Tell me, Slade," I said, after we were seated, "is this doctrine of yours new or are you continuing a long tradition?"

"It's by no means new," he replied, "though I *can* claim to have established the first church in which hyperfaith serves as the cornerstone. Plato, of course, had no conception of a geometrical fourth dimension, though his cave analogy clearly implies it. In fact, every form of Platonic dualism that divides existence into the natural and supernatural is clearly a nonmathematical way of speaking about higher space. Henry More, the seventeenth-century Cambridge Platonist, was the first to regard the spiritual world as having four spatial dimensions. Then along came Immanuel Kant, with his recognition of our space and time as subjective lenses, so to speak, through which we view only a thin slice of transcendent reality. After that it is easy to see how the concept of higher space provided a much needed link between modern science and traditional religions."

"You say 'religions,'" I put in. "Does that mean your church is not Christian?"

"Only in the sense that we find essential truth in all the great world faiths. I should add that in recent decades the Continental Protestant theologians have finally discovered four-space. When Karl Barth talks about the 'vertical' or 'perpendicular' dimension, he clearly means it in a four-dimensional sense. And of course in the theology of Karl Heim there is a full, explicit recognition of the role of higher space."

"Yes," I said. "I recently read an interesting book called *Physicist and Christian*, by William G. Pollard [executive director of the Oak Ridge Institute of Nuclear Studies and an Episcopal clergyman]. He draws heavily on Heim's concept of hyperspace."

Slade scribbled the book's title on a note pad. "I must look it up. I wonder if Pollard realizes that a number of late-nineteenth-century Protestants wrote books about the fourth dimension. A. T. Schofield's *Another World*, for example [it appeared in 1888], and Arthur Willink's *The World of the Unseen* [subtitled "An Essay on the Relation of Higher Space to Things Eternal"; published in 1893]. Of course modern occultists and spiritualists have had a field day with the notion. Peter D. Ouspensky, for instance, has a lot to say about it in his books, although most of his opinions derive from the speculations of Charles Howard Hinton, an American mathematician. Whately Carington, the English parapsychologist, wrote an unusual book in 1920—he published it under the by-line of W. Whately Smith—on *A Theory of the Mechanism of Survival*."

"Survival after death?"

Slade nodded. "I can't go along with Carington's belief in such things as table tipping being accomplished by an invisible four-dimensional lever, or clairvoyance as perception from a point in higher space, but I regard his basic hypothesis as sound. Our bodies are simply three-dimensional cross sections of our higher four-dimensional selves. Obviously a man is subject to all the laws of this world, but at the same time his experiences are permanently recorded—stored as information, so to speak—in the four-space portion of his higher self. When his three-space body ceases to function, the permanent record remains until it can be attached to a new body for a new cycle of life in some other three-space continuum."

"I like that," I said. "It explains the complete dependence of mind on body in this world, at the same time permitting an unbroken continuity between this life and the next. Isn't this close to what William James struggled to say in his little book on immortality?"

"Precisely. James, unfortunately, was no mathematician, so he had to express his meaning in nongeometrical metaphors."

"What about the so-called demonstrations of the fourth dimension by certain mediums," I asked. "Wasn't there a professor of astrophysics in Leipzig who wrote a book about them?"

I thought I detected an embarrassed note in Slade's laugh. "Yes, that was poor Johann Karl Friedrich Zöllner. His book *Transcendental Physics* was translated into English in 1881, but even the English copies are now quite rare. Zöllner did some good work in spectrum analysis, but he was supremely ignorant of conjuring methods. As a consequence he was badly taken in, I'm afraid, by Henry Slade, the American medium."

"Slade?" I said with surprise.

"Yes, I'm ashamed to say we're related. He was my great-uncle. When he died, he left a dozen fat notebooks in which he had recorded his methods. Those notebooks were acquired by the English side of my family and handed down to me."

"This excites me greatly," I said. "Can you demonstrate any of the tricks?"

The request seemed to please him. Conjuring, he explained, was one of his hobbies, and he thought that the mathematical angles of several of Henry's tricks would be of interest to my readers.

From a drawer in his desk Slade took a strip of leather, cut as shown at the left in Figure 26, to make three parallel

Figure 26
Slade's leather strip—braided
in hyperspace?

strips. He handed me a ball-point pen with the request that I mark the leather in some way to prevent later substitution. I initialed a corner as shown. We sat on opposite sides of a small table. Slade held the leather under the table for a few moments, then brought it into view again. It was braided exactly as shown at the right in the illustration! Such braiding would be easy to accomplish if one could move the strips through hyperspace. In three-space it seemed impossible.

Slade's second trick was even more astonishing. He had me examine a rubber band of the wide, flat type shown at the left in Figure 27. This was placed in a matchbox, and the box

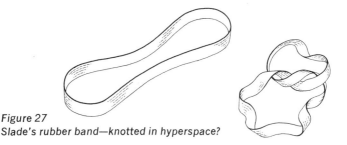

Figure 27
Slade's rubber band—knotted in hyperspace?

was securely sealed at both ends with cellophane tape. Slade started to place it under the table, then remembered he had forgotten to have me mark the box for later identification. I drew a heavy X on the upper surface.

"If you like," he said, "you yourself may hold the box under the table."

I did as directed. Slade reached down, taking the box by its other end. There was a sound of movement and I could feel that the box seemed to be vibrating slightly.

Slade released his grip. "Please open the box."

First I inspected the box carefully. The tape was still in place. My mark was on the cover. I slit the tape with my thumbnail and pushed open the drawer. The elastic band—*mirabile dictu*—was tied in a simple knot as shown at the right in Figure 27.

"Even if you managed somehow to open the box and switch bands," I said, "how the devil could you get a rubber band like this?"

Slade chuckled. "My great-uncle was a clever rascal."

I was unable to persuade Slade to tell me how either trick

was done. The reader is invited to think about them before he reads this chapter's answer section.

We talked of many other things. When I finally left the Church of the Fourth Dimension, a heavy fog was swirling through the wet streets of London. I was back in Plato's cave. The shadowy forms of moving cars, their headlights forming flat elliptical blobs of light, made me think of some familiar lines from the Rubáiyát of a great Persian mathematician:

> *We are no other than a moving row*
> *Of magic shadow-shapes that come and go*
> *Round with the sun-illumined lantern held*
> *In midnight by the Master of the Show.*

ADDENDUM

Although I spoke in the first paragraph of this chapter of an "imaginary visit" to London, when the chapter first appeared in *Scientific American* several readers wrote to ask for the address of Slade's church. The Reverend Slade is purely fictional, but Henry Slade the medium was one of the most colorful and successful mountebanks in the history of American spiritualism. I have written briefly about him and given the major references in a chapter on the fourth dimension in my book *The Ambidextrous Universe* (New York: Basic Books, 1964; London: Allen Lane, 1967).

ANSWERS

Slade's method of braiding the leather strip is familiar to Boy Scouts in England and to all those who make a hobby of leathercraft. Many readers wrote to tell me of books in which this type of braiding is described: George Russell Shaw, *Knots, Useful and Ornamental* (page 86); Constantine A. Belash, *Braiding and Knotting* (page 94); Clifford Pyle, *Leather Craft as a Hobby* (page 82); Clifford W. Ashley, *The Ashley Book of Knots* (page 486); and others. For a full mathematical analysis, see J. A. H. Shepperd, "Braids Which Can Be Plaited with Their Threads Tied Together at Each End," *Proceedings of the Royal Society*, A, Vol. 265 (1962), pages 229–44.

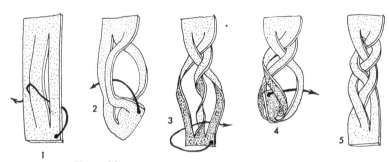

Figure 28
Slade's first trick

There are several ways to go about making the braid. Figure 28 was drawn by reader George T. Rab of Dayton, Ohio. By repeating this procedure one can extend the braid to any multiple of six crossings. Another procedure is simply to form the six-cross plat in the upper half of the strip by braiding in the usual manner. This creates a mirror image of the plat in the lower half. The lower plat is easily removed by one hand while the upper plat is held firmly by the other hand. Both procedures can be adapted to leather strips with more than three strands. If stiff leather is used, it can be made pliable by soaking it in warm water.

Slade's trick of producing a knot in a flat rubber band calls first for the preparation of a knotted band. Obtain a rubber ring of circular cross section and carefully carve a portion of it flat as shown in Figure 29. Make three half twists in the flat section (*middle drawing*), then continue carving the rest of the ring to make a flat band with three half twists (*bottom drawing*). Mel Stover of Winnipeg, Canada, suggests that this can best be done by stretching the ring around a wooden block, freezing the ring, then flattening it with a home grind-

Figure 29
Slade's second trick

ing tool. When the final band is cut in half all the way around, it forms a band twice as large and tied in a single knot.

A duplicate band of the same size, but unknotted, must also be obtained. The knotted band is placed in a matchbox and the ends of the box are sealed with tape. It is now necessary to substitute this matchbox for the one containing the unknotted band. I suspect that Slade did this when he started to put the box under the table, then "remembered" that I had not yet initialed it. The prepared box could have been stuck to the underside of the table with magician's wax. It would require only a moment to press the unprepared box against another dab of wax, then take the prepared one. In this way the switch occurred *before* I marked the box. The vibrations I felt when Slade and I held the box under the table were probably produced by one of Slade's fingers pressing firmly against the box and sliding across it.

Fitch Cheney, mathematician and magician, wrote to tell about a second and simpler way to create a knotted elastic band. Obtain a hollow rubber torus—they are often sold as teething rings for babies—and cut as shown by the dotted line in Figure 30. The result is a wide endless band tied in a single knot. The band can be trimmed, of course, to narrower width.

It was Stover, by the way, who first suggested to me the problem of tying a knot in an elastic band. He had been shown such a knotted band by magician Winston Freer. Freer said he knew *three* ways of doing it.

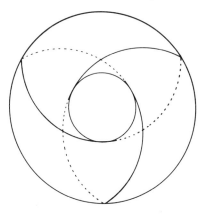

Figure 30
A second way to produce a
knotted rubber band

CHAPTER SEVEN

O

Eight Problems

1. A Digit-Placing Problem

THIS PERPLEXING DIGITAL PROBLEM, inventor unknown, was passed on to me by L. Vosburgh Lyons of New York City. The digits from 1 to 8 are to be placed in the eight circles shown in Figure 31, with this proviso: no two digits directly

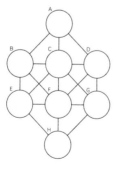

Figure 31
A perplexing digital problem

adjacent to each other in serial order may go in circles that are directly connected by a line. For example, if 5 is placed in the top circle, neither 4 nor 6 may be placed in any of the three circles that form a horizontal row beneath it, because

each of these circles is joined directly to the top circle by a
straight line.

There is only one solution (not counting a rotation or mirror reflection as being different), but if you try to find it
without a logical procedure, the task will be difficult.

2. The Lady or the Tiger?

FRANK STOCKTON'S famous short story "The Lady or the
Tiger?" tells of a semi-barbaric king who enjoyed administering a curious kind of justice. The king sat on a high
throne at one side of his public arena. On the opposite side
were twin doors. The prisoner on trial could open either
door, guided only by "impartial and incorruptible chance."
Behind one door was a hungry tiger; behind the other, a desirable young lady. If the tiger sprang through the door, the
man's fate was considered a just punishment for his crime.
If the lady stepped forth, the man's innocence was rewarded
by a marriage ceremony performed on the spot.

The king, having discovered his daughter's romance with a
certain courtier, has placed the unfortunate young man on
trial. The princess knows which door conceals the tiger. She
also knows that behind the other door is the fairest lady of
the court, whom she has observed making eyes at her lover.
The courtier knows the princess knows. She makes a "slight,
quick movement" of her hand to the right. He opens the door
on the right. The tale closes with the tantalizing question:
"Which came out of the opened door—the lady or the tiger?"

After extensive research on this incident, I am able to
make the first full report on what happened next. The two
doors were side by side and hinged to open toward each
other. After opening the door on the right the courtier
quickly pulled open the other door and barricaded himself
inside the triangle formed by the doors and the wall. The
tiger emerged through one door, entered the other and ate
the lady.

The king was a bit nonplused, but, being a good sport, he
allowed the courtier a second trial. Not wishing to give the
wily young man another 50–50 chance, he had the arena reconstructed so that instead of one pair of doors there were

now three pairs. Behind one pair he placed two hungry tigers. Behind the second pair he placed a tiger and a lady. Behind the third pair he placed two ladies who were identical twins and who were dressed exactly alike.

The cruel scheme was as follows. The courtier must first choose a pair of doors. Then he must select one of the two and a key would be tossed to him for opening it. If the tiger emerged, that was that. If the lady, the door would immediately be slammed shut. The lady and her unknown partner (either her twin sister or a tiger) would then be secretly rearranged in the same two rooms, one to a room, according to a flip of a special gold coin with a lady on one side and a tiger on the other. The courtier would be given a second choice between the same two doors, without knowing whether the arrangement was different or the same as before. If he chose a tiger, that was that again; if a lady, the door would be slammed shut, the coin-flipping procedure repeated to determine who went in which room, and the courtier given a third and final choice of one of the same two doors. If successful in his last choice, he would marry the lady and his ordeal would be over.

The day of the trial arrived and all went according to plan. Twice the courtier selected a lady. He tried his best to determine if the second lady was the same as the first but was unable to decide. Beads of perspiration glistened on his forehead. The face of the princess—she was ignorant this time of who went where—was as pale as white marble.

Exactly what probability did the courtier have of finding a lady on his third guess?

3. A Tennis Match

MIRANDA beat Rosemary in a set of tennis, winning six games to Rosemary's three. Five games were won by the player who did not serve. Who served first?

4. The Colored Bowling Pins

A WEALTHY MAN had two bowling lanes in his basement. In one lane ten dark-colored pins were used; in the other, ten light-colored pins. The man had a mathematical turn of mind, and the following problem occurred to him one evening as he was practicing his delivery:

Is it possible to mix pins of both colors, then select ten pins that can be placed in the usual triangular formation in such a way that no three pins of the same color will mark the vertices of an equilateral triangle?

If it is possible, show how to do it. Otherwise prove that it cannot be done. A set of checkers will provide convenient pieces for working on the problem.

5. The Problem of the Six Matches

PROFESSOR LUCIUS S. WILSUN is a brilliant, though somewhat eccentric, topologist. His name had formerly been Wilson. As a graduate student he had noted that when his full name, Lucius Sims Wilson, was printed in capital letters, all the letters were topologically equivalent except for the O. This so annoyed him that he had his name legally changed.

When I met him for lunch recently, I found him forming patterns on the tablecloth with six paper matches. "A new topological puzzle?" I asked hopefully.

"In a way," he replied. "I'm trying to find out how many topologically distinct patterns I can make with six matches by placing them flat on the table, without crossing one match over another, and joining them only at the ends."

"That shouldn't be difficult," I said.

"Well, it's trickier than you might think. I've just worked out all the patterns for smaller numbers of matches." He handed me an envelope on the back of which he had jotted down a rough version of the chart in Figure 32.

"Didn't you overlook a five-match pattern?" I said. "Consider that third figure—the square with the tail. Suppose you put the tail *inside* the square. If the matches are confined to the plane, obviously one pattern can't be deformed into the other."

Wilsun shook his head. "That's a common misconception about topological equivalence. It's true that if one figure can be changed to another by pulling and stretching, without breaking or tearing, the two must be topologically identical

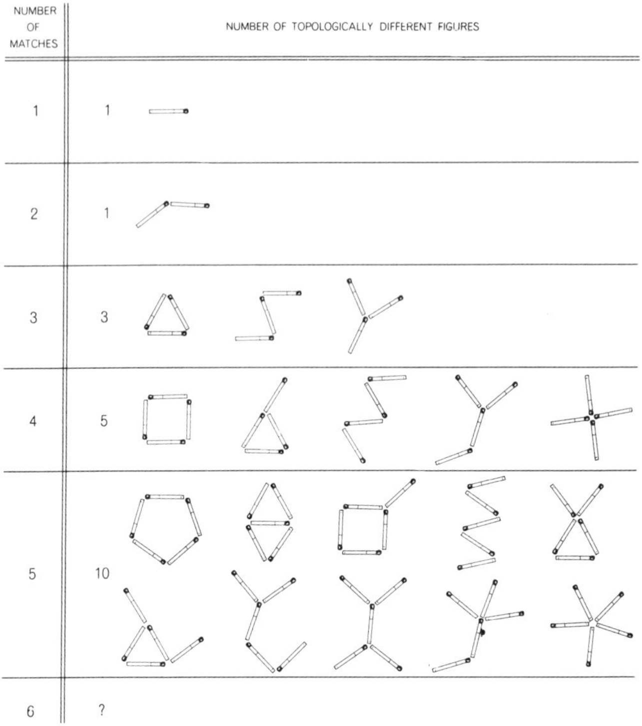

Figure 32
A chart of the topologically distinct patterns that can be made with one to six matches

—as we topologists like to say, homeomorphic. But not the other way around. If two figures are homeomorphic, it is *not* always possible to deform one into the other."

"I beg your pardon," I said.

"Don't topologize. Two figures are homeomorphic if, as you move continuously from point to point along one figure, you can make a corresponding movement from point to point —the points of the two figures must be in one-to-one correspondence, of course—along the other figure. For example, a piece of rope joined at the ends is homeomorphic with a piece of rope that is knotted before the ends are joined, although you obviously cannot deform one to the other. Two spheres that touch externally are homeomorphic with two spheres of different size, the smaller inside the other and touching at one point."

I must have looked puzzled, because he quickly added: "Look, here's a simple way to make it clear to your readers. Those match figures are on the plane, but think of them as elastic bands. You can pick them up, manipulate them any way you wish, turn them over if you please, put them back down again. If one figure can be changed to another this way, they are topologically the same."

"I see," I said. "If you think of a figure as embedded in a higher space, then it *is* possible to deform one figure into any other figure that is topologically equivalent to it."

"Precisely. Imagine the endless rope or the two spheres in a four-dimensional space. The knot can be tied or untied while the ends remain joined. The small sphere can be moved in or out of the larger one."

With this understanding of topological equivalence, the reader is asked to determine the exact number of topologically different figures that can be formed on the plane with six matches. Remember, the matches themselves are rigid and all the same size. They must not be bent or stretched, they must not overlap and they may touch only at their ends. But once a figure is formed it must be thought of as an elastic structure that can be picked up, deformed in three-space, then returned to the plane. The figures are not graphs in which vertices, where two matches join, keep their identity. Thus a triangle is equivalent to a square or a pentagon; a chain of two matches is equivalent to a chain of any length; the capital letters E, F, T and Y are all equivalent; R is the same as its mirror image; and so on.

6. Two Chess Problems: Minimum and Maximum Attacks

MANY BEAUTIFUL CHESS PROBLEMS do not involve positions of competitive play; they use the pieces and board only for posing a challenging mathematical task. Here are two classic task problems that surely belong together:

1. The minimum-attack problem: Place the eight pieces of one color (king, queen, two bishops, two knights, two rooks) on the board so that the *smallest* possible number of squares are under attack. A piece does not attack the square on which it rests, but of course it may attack squares occupied

by other pieces. In Figure 33, twenty-two squares (gray) are under attack, but this number can be reduced considerably. It is not necessary that the two bishops be placed on opposite colors.

2. The maximum-attack problem: Place the same eight pieces on the board so that the *largest* possible number of squares are under attack. Again, a piece does not attack its own square, but it may attack other occupied squares. The two bishops need not be on opposite colors. In Figure 34, fifty-five squares (gray) are under attack. This is far from the maximum.

There is a proof for the maximum number when the bishops are on squares of the same color. No one has yet proved the maximum when the bishops are on different colors. The minimum is believed to be the same regardless of whether the bishops are on the same or opposite colors, but both cases are unsupported by proof. So many chess experts have worked on these problems that it is not likely any of the conjectured answers will be modified. Should any reader beat the records, it will be big news in chess-problem circles.

7. How Far Did the Smiths Travel?

AT TEN O'CLOCK ONE MORNING Mr. Smith and his wife left their house in Connecticut to drive to the home of Mrs. Smith's parents in Pennsylvania. They planned to stop once

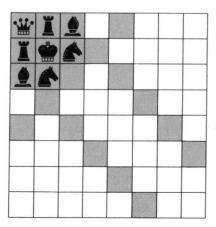

Figure 33
The minimum-attack problem

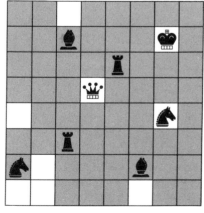

Figure 34
The maximum-attack problem

along the way for lunch at Patricia Murphy's Candlelight Restaurant in Westchester.

The prospective visit with his in-laws, combined with business worries, put Mr. Smith in a sullen, uncommunicative mood. It was not until eleven o'clock that Mrs. Smith ventured to ask: "How far have we gone, dear?"

Mr. Smith glanced at the mileage meter. "Half as far as the distance from here to Patricia Murphy's," he snapped.

They arrived at the restaurant at noon, enjoyed a leisurely lunch, then continued on their way. Not until five o'clock, when they were 200 miles from the place where Mrs. Smith had asked her first question, did she ask a second one. "How much farther do we have to go, dear?"

"Half as far," he grunted, "as the distance from here to Patricia Murphy's."

They arrived at their destination at seven that evening. Because of traffic conditions Mr. Smith had driven at widely varying speeds. Nevertheless, it is quite simple to determine (and this is the problem) exactly how far the Smiths traveled from one house to the other.

8. Predicting a Finger Count

ON LAST NEW YEAR'S DAY a mathematician was puzzled by the strange way in which his small daughter began to count on the fingers of her left hand. She started by calling the thumb 1, the first finger 2, middle finger 3, ring finger 4, little finger 5, then she reversed direction, calling the ring finger 6, middle finger 7, first finger 8, thumb 9, then back to the first finger for 10, middle finger for 11, and so on. She continued to count back and forth in this peculiar manner until she reached a count of 20 on her ring finger.

"What in the world are you doing?" her father asked.

The girl stamped her foot. "Now you've made me forget where I was. I'll have to start all over again. I'm counting up to 1962 to see what finger I'll end on."

The mathematician closed his eyes while he made a simple mental calculation. "You'll end on your ——," he said.

When the girl finished her count and found that her father was right, she was so impressed by the predictive power of mathematics that she decided to work twice as hard on her arithmetic lessons. How did the father arrive at his prediction and what finger did he predict?

ANSWERS

1. If the numbers from 1 to 8 are placed in the circles as shown in Figure 35, no number will be connected by a line to a number immediately above or below it in serial order.

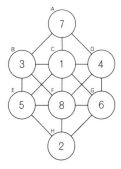

Figure 35
Solution to Problem 1

The solution (including its upside-down and mirror-image forms) is unique.

L. Vosburgh Lyons solved it as follows. In the series 1, 2, 3, 4, 5, 6, 7, 8 each digit has two neighboring numbers except 1 and 8. In the diagram, circle C is connected to every circle except H. Therefore if C contains any number in the set 2, 3, 4, 5, 6, 7, only circle H will remain to accommodate *both* neighbors of whatever number goes in C. This is impossible, so C must contain 1 or 8. The same argument applies to circle F. Because of the pattern's symmetry, it does not matter whether 1 goes in C or F, so let us place it in C. Circle H is the only circle available for 2. Similarly, with 8 in circle F, only circle A is available for 7. The remaining four numbers are now easily placed.

Thomas H. O'Beirne, Glasgow, and Herb Koplowitz, Elmont, New York, each solved the problem by drawing a new diagram in which the old network of lines is replaced by lines connecting all circles that were *not* connected before.

The original problem now takes the form of placing the digits in the circles so that a connected path can be traced from 1 to 8, taking the digits in order. It is easy to see, by inspection of the new diagram, that only four ways of placing the digits are possible, and they correspond to the rotations and reflections of the unique solution.

Fred Gruenberger of the Rand Corporation, Santa Monica, wrote to say that he had encountered this problem about a year earlier "through a friend at the Walt Disney Studios, where it had already consumed a fair amount of Mr. Disney's staff time." Gruenberger had used it as the basis of a West Coast television show, "How a Digital Computer Works," to illustrate the difference between the way a human mathematician approaches such a problem and the brute-force approach of a digital computer that finds the solution by running through all possible permutations of the digits, in this case 40,320 different arrangements.

2. The problem of the lady or the tiger is merely a dressed-up version of a famous ball-and-urn problem analyzed by the great French mathematician Pierre Simon de Laplace (see James R. Newman's *The World of Mathematics* [New York: Simon and Schuster, 1956], Vol. 2, page 1332). The answer is that the young man on his third choice of a door has a probability of 9/10 that he will choose the lady. The pair of doors concealing two tigers is eliminated by his first choice of a lady, which leaves 10 equally probable possibilities for the entire series of three choices.

If the doors conceal two ladies:
Lady 1 — Lady 1 — Lady 1
Lady 1 — Lady 1 — Lady 2
Lady 1 — Lady 2 — Lady 1
Lady 1 — Lady 2 — Lady 2
Lady 2 — Lady 1 — Lady 1
Lady 2 — Lady 1 — Lady 2
Lady 2 — Lady 2 — Lady 1
Lady 2 — Lady 2 — Lady 2
If the doors conceal a lady and a tiger:
Lady 3 — Lady 3 — Lady 3
Lady 3 — Lady 3 — Tiger
Of the 10 possibilities in the problem's "sample space,"

only one ends with a fatal final choice. The probability of the man's survival is therefore 9/10.

3. The solution I gave in *Scientific American* was so long and awkward that many readers provided shorter and better ones. W. B. Hogan and Paul Carnahan each found simple algebraic solutions, Peter M. Addis and Martin T. Pett each made use of a simple diagram, and Thomas B. Gray, Jr., solved the problem by an ingenious use of the binary system. The shortest solution came from Goran Ohlin, an economist at Columbia University, who expressed it as follows:

"Whoever served first, served five games, and the other player served four. Suppose the first server won x of the games she served and y of the other four games. The total number of games *lost* by the player who served them is then $5 - x + y$. This equals 5 [we were told that the non-server won five games], therefore $x = y$, and the first player won a total of $2x$ games. Because only Miranda won an even number of games, she must have been the first server."

4. It is not possible to mix bowling pins of two different colors and set up a triangular formation of ten pins in such a way that no three pins of the same color mark the corners of an equilateral triangle. There are many ways to prove this. The following is typical:

Assume that the two colors are red and black and that the 5 pin (*see Figure 36*) is red. Pins 4, 9, 3 form an equilateral triangle, so at least one of these pins must be red. It does not matter which we make red, because of the figure's symmetry, so let us make it the 3 pin. Pins 2 and 6 must therefore be black. Pins 2, 6, 8 form a triangle, forcing us to make 8 a red pin. This in turn makes the 4 and 9 pins black. Pin 10 cannot be black, for this would form a black triangle with 6 and 9, nor can it be red, because this would form a red triangle with 3 and 8. Therefore pin 5, with which we start-

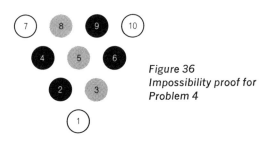

Figure 36
Impossibility proof for
Problem 4

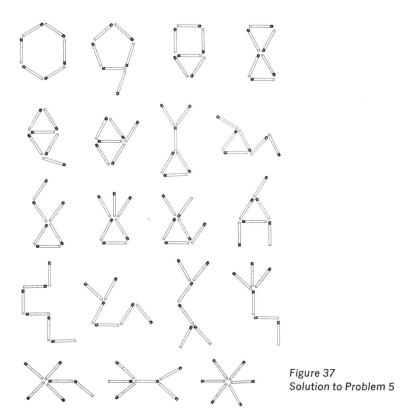

Figure 37
Solution to Problem 5

ed, cannot be red. Of course, the same argument will show that it cannot be black.

5. Nineteen topologically distinct networks can be made with six matches, placing them on a plane so that no matches overlap and the matches touch only at their ends. The nineteen networks are shown in Figure 37. If the restriction to a plane is dropped and three-space networks permitted, only one additional figure is possible: the skeleton of a tetrahedron.

Readers William G. Hoover and Victoria N. Hoover, Durham, North Carolina, Ronald Read, University of London, and Henry Eckhardt, Fair Oaks, California, extended the problem to seven matches and found thirty-nine topologically distinct patterns.

6. Figure 38 shows how eight chess pieces of one color can be placed on the board so that only sixteen squares are under attack. The queen and bishop in the corner can be switched

to provide a 16-square minimum with bishops of the same color. This is believed to be the minimum regardless of whether the bishops are the same color or different colors. The position also solves two other minimum problems for the eight pieces: a minimum number of moves (ten), and a minimum number of pieces (three) that can move.

Figure 39 shows one way to place the eight pieces so that all 64 squares are under attack, obviously the maximum. With bishops of opposite color, 63 is believed to be the maximum. There are scores of distinct solutions, one of which is shown in Figure 40. The exact number of different solutions is not known.

The maximum-attack problem with bishops of opposite color was first proposed by J. Kling in 1849 with the added proviso that the king occupy the single unattacked square. Readers may enjoy searching for such a solution, as well as for a pattern (unusually difficult) in which the unattacked square is at the corner of the board. Two readers, C. C. Verbeek, The Hague, and Roger Maddux, Arcadia, California, sent identical solutions of the second problem, with the unattacked corner square occupied by a rook. It has been shown that the unattacked square may be at any spot on the board.

7. To find the mileage covered by the Smiths on their trip from Connecticut to Pennsylvania, the various times of day that are given are irrelevant, since Smith drove at varying speeds. At two points along the way Mrs. Smith asked a question. Smith's answers indicate that the distance from the

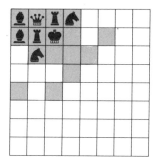

*Figure 38
Solution to minimum-attack problem*

*Figure 39
Solution to maximum-attack problem with bishops on same color*

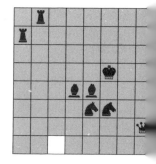

*Figure 40
Solution to maximum-attack problem with bishops on different colors*

first point to Patricia Murphy's Candlelight Restaurant is two thirds of the distance from the start of the trip to the restaurant, and the distance from the restaurant to the second point is two thirds of the distance from the restaurant to the end of the trip. It is obvious, therefore, that the distance from point to point (which we are told is 200 miles) is two thirds of the total distance. This makes the total distance 300 miles. Figure 41 should make it all clear.

Figure 41
Chart for Problem 7

200 MILES

QUESTION 1   RESTAURANT   QUESTION 2

8. When the mathematician's little girl counted to 1,962 on her fingers, counting back and forth in the manner described, the count ended on her index finger. The fingers are counted in repetitions of a cycle of eight counts as shown in Figure 42. It is a simple matter to apply the concept of numerical congruence, modulo 8, in order to calculate where the count will fall for any given number. We have only to divide the number by 8, note the remainder, then check to see which finger is so labeled. The number 1,962 divided by 8 has a remainder of 2, so the count falls on the index finger.

In mentally dividing 1,962 by 8 the mathematician recalled the rule that any number is evenly divisible by 8 if its last three digits are evenly divisible by 8, so he had only to divide 962 by 8 to determine the remainder.

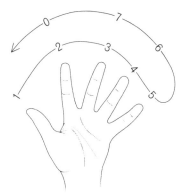

Figure 42
How fingers are labeled for
Problem 8

CHAPTER EIGHT

○

A Matchbox
Game-Learning
Machine

I knew little of chess, but as only a few pieces were on the board, it was obvious that the game was near its close. . . . [Moxon's] face was ghastly white, and his eyes glittered like diamonds. Of his antagonist I had only a back view, but that was sufficient; I should not have cared to see his face.

THE QUOTATION is from Ambrose Bierce's classic robot story, "Moxon's Master" (reprinted in Groff Conklin's excellent science-fiction anthology, *Thinking Machines*). The inventor Moxon has constructed a chess-playing robot. Moxon wins a game. The robot strangles him.

Bierce's story reflects a growing fear. Will computers someday get out of hand and develop a will of their own? Let it not be thought that this question is asked today only by those who do not understand computers. Before his death Norbert Wiener anticipated with increasing apprehension the day when complex government decisions would be turned over to sophisticated game-theory machines. Before we know it, Wiener warned, the machines may shove us over the brink into a suicidal war.

90

The greatest threat of unpredictable behavior comes from the learning machines: computers that improve with experience. Such machines do not do what they have been told to do but what they have *learned* to do. They quickly reach a point at which the programmer no longer knows what kinds of circuits his machine contains. Inside most of these computers are randomizing devices. If the device is based on the random decay of atoms in a sample radioactive material, the machine's behavior is not (most physicists believe) predictable even in principle.

Much of the current research on learning machines has to do with computers that steadily improve their ability to play games. Some of the work is secret—war is a game. The first significant machine of this type was an IBM 704 computer programed by Arthur L. Samuel of the IBM research department at Poughkeepsie, New York. In 1959 Samuel set up the computer so that it not only played a fair game of checkers but also was capable of looking over its past games and modifying its strategy in the light of this experience. At first Samuel found it easy to beat his machine. Instead of strangling him, the machine improved rapidly, soon reaching the point at which it could clobber its inventor in every game. So far as I know no similar program has yet been designed for chess, although there have been several ingenious programs for nonlearning chess machines.

A few years ago the Russian chess grandmaster Mikhail Botvinnik was quoted as saying that the day would come when a computer would play master chess. "This is of course nonsense," wrote the American chess expert Edward Lasker in an article on chess machines in the Fall 1961 issue of a magazine called *The American Chess Quarterly*. But it was Lasker who was talking nonsense. A chess computer has three enormous advantages over a human opponent: (1) it never makes a careless mistake; (2) it can analyze moves ahead at a speed much faster than a human player can; (3) it can improve its skill without limit. There is every reason to expect that a chess-learning machine, after playing thousands of games with experts, will someday develop the skill of a master. It is even possible to program a chess machine to play continuously and furiously against itself.

Its speed would enable it to acquire in a short time an experience far beyond that of any human player.

It is not necessary for the reader who would like to experiment with game-learning machines to buy an electronic computer. It is only necessary to obtain a supply of empty matchboxes and colored beads. This method of building a simple learning machine is the happy invention of Donald Michie, a biologist at the University of Edinburgh. Writing on "Trial and Error" in *Penguin Science Survey 1961,* Vol. 2, Michie describes a ticktacktoe learning machine called MENACE (Matchbox Educable Naughts And Crosses Engine) that he constructed with three hundred matchboxes.

MENACE is delightfully simple in operation. On each box is pasted a drawing of a possible ticktacktoe position. The machine always makes the first move, so only patterns that confront the machine on odd moves are required. Inside each box are small glass beads of various colors, each color indicating a possible machine play. A V-shaped cardboard fence is glued to the bottom of each box, so that when one shakes the box and tilts it, the beads roll into the V. Chance determines the color of the bead that rolls into the V's corner. First-move boxes contain four beads of each color, third-move boxes contain three beads of each color, fifth-move boxes have two beads of each color, seventh-move boxes have single beads of each color.

The robot's move is determined by shaking and tilting a box, opening the drawer and noting the color of the "apical" bead (the bead in the V's apex). Boxes involved in a game are left open until the game ends. If the machine wins, it is rewarded by adding three beads of the apical color to each open box. If the game is a draw, the reward is one bead per box. If the machine loses, it is punished by extracting the apical bead from each open box. This system of reward and punishment closely parallels the way in which animals and even humans are taught and disciplined. It is obvious that the more games MENACE plays, the more it will tend to adopt winning lines of play and shun losing lines. This makes it a legitimate learning machine, although of an extremely simple sort. It does not make (as does Samuel's checker machine) any self-analysis of past plays that causes it to devise new strategies.

Michie's first tournament with MENACE consisted of 220 games over a two-day period. At first the machine was easily trounced. After seventeen games the machine had abandoned all openings except the corner opening. After the twentieth game it was drawing consistently, so Michie began trying unsound variations in the hope of trapping it in a defeat. This paid off until the machine learned to cope with all such variations. When Michie withdrew from the contest after losing eight out of ten games, MENACE had become a master player.

Since few readers are likely to attempt building a learning machine that requires three hundred matchboxes, I have designed hexapawn, a much simpler game that requires only twenty-four boxes. The game is easily analyzed—indeed, it is trivial—but the reader is urged *not* to analyze it. It is much more fun to build the machine, then learn to play the game while the machine is also learning.

Hexapawn is played on a 3 × 3 board, with three chess pawns on each side as shown in Figure 43. Dimes and pennies can be used instead of actual chess pieces. Only two types of move are allowed: (1) A pawn may advance straight forward one square to an empty square; (2) a pawn may capture an enemy pawn by moving one square diagonally, left or right, to a square occupied by the enemy. The captured piece is removed from the board. These are the same as pawn moves in chess, except that no double move, *en passant* capture or promotion of pawns is permitted.

The game is won in any of three ways:

1. By advancing a pawn to the third row.

2. By capturing all enemy pieces.

3. By achieving a position in which the enemy cannot move.

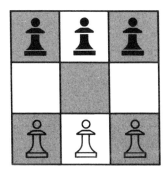

Figure 43
The game of hexapawn

Players alternate moves, moving one piece at a time. A draw clearly is impossible, but it is not immediately apparent whether the first or second player has the advantage.

To construct HER (Hexapawn Educable Robot) you need twenty-four empty matchboxes and a supply of colored beads. Small candies that come in different colors—jujubes for example—or colored popping corn also work nicely. Each matchbox bears one of the diagrams in Figure 44. The robot always makes the second move. Patterns marked "2" represent the two positions open to HER on the second move. You have a choice between a center or an end opening, but only the left end is considered because an opening on the right would obviously lead to identical (although mirror-reflected) lines of play. Patterns marked "4" show the eleven positions that can confront HER on the fourth (its second) move. Patterns marked "6" are the eleven positions that can face HER on the sixth (its last) move. (I have included mirror-image patterns in these positions to make the working easier; otherwise nineteen boxes would suffice.)

Inside each box place a single bead to match the color of each arrow on the pattern. The robot is now ready for play. Every legal move is represented by an arrow; the robot can therefore make all possible moves and only legal moves. The robot has no strategy. In fact, it is an idiot.

The teaching procedure is as follows. Make your first move. Pick up the matchbox that shows the position on the board. Shake the matchbox, close your eyes, open the drawer, remove one bead. Close the drawer, put down the box, place the bead on top of the box. Open your eyes, note the color of the bead, find the matching arrow and move accordingly. Now it is your turn to move again. Continue this procedure until the game ends. If the robot wins, replace all the beads and play again. If it loses, punish it by confiscating only the bead that represents its *last* move. Replace the other beads and play again. If you should find an empty box (this rarely happens), it means the machine has no move that is not fatal and it resigns. In this case confiscate the bead of the preceding move.

Keep a record of wins and losses so you can chart the first fifty games. Figure 45 shows the results of a typical fifty-

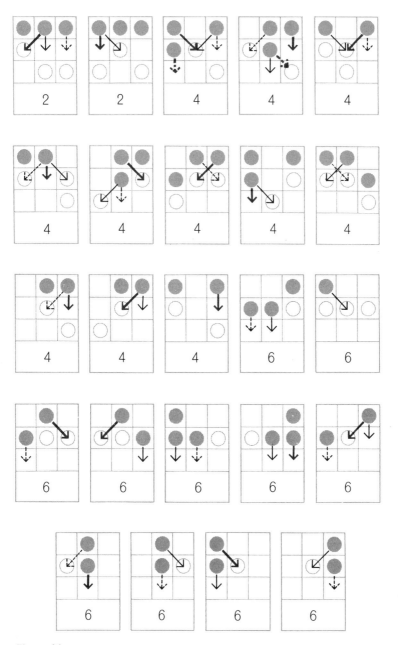

Figure 44
Labels for HER matchboxes. (The four different kinds of arrows represent four different colors.)

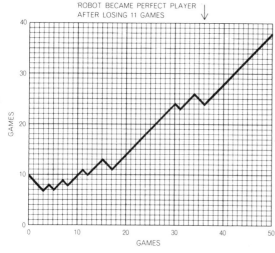

Figure 45
Learning curve for HER's first
fifty games (downslant shows
loss, upslant a win)

game tournament. After thirty-six games (including eleven defeats for the robot) it has learned to play a perfect game. The system of punishment is designed to minimize the time required to learn a perfect game, but the time varies with the skill of the machine's opponent. The better the opponent, the faster the machine learns.

The robot can be designed in other ways. For example, if the intent is to maximize the number of games that the machine wins in a tournament of, say, twenty-five games, it may be best to reward (as well as punish) by adding a bead of the proper color to each box when the machine wins. Bad moves would not be eliminated so rapidly, but it would be less inclined to make the bad moves. An interesting project would be to construct a second robot, HIM (Hexapawn Instructable Matchboxes), designed with a different system of reward and punishment but equally incompetent at the start of a tournament. Both machines would have to be enlarged so they could make either first or second moves. A tournament could then be played between HIM and HER, alternating the first move, to see which machine would win the most games out of fifty.

Similar robots are easily built for other games. Stuart C. Hight, director of research studies at the Bell Telephone Laboratories in Whippany, New Jersey, recently built a matchbox learning machine called NIMBLE (Nim Box Logic Engine) for playing Nim with three piles of three counters each. The robot plays either first or second and is rewarded or punished after each game. NIMBLE required only eighteen matchboxes and played almost perfectly after thirty games. For an analysis of the game of Nim, see Chapter 15 of my *Scientific American Book of Mathematical Puzzles & Diversions.*

By reducing the size of the board the complexity of many familiar games can be minimized until they are within the scope of a matchbox robot. The game of go, for example, can be played on the intersections of a 2 × 2 checkerboard. The smallest nontrivial board for checkers is shown in Figure 46. It should not be difficult to build a matchbox machine that would learn to play it. Readers disinclined to do this may enjoy analyzing the game. Does either side have a sure win or will two perfect players draw?

When chess is reduced to the smallest board on which all legal moves are still possible, as shown in Figure 46, the complexity is still far beyond the capacity of a matchbox machine. In fact, I have found it impossible to determine which player, if either, has the advantage. Minichess is recommended for computer experts who wish to program a simplified chess-learning machine and for all chess players who like to sneak in a quick game during a coffee break.

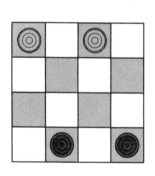 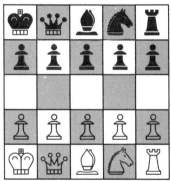

Figure 46
Matchbox machines can be built for minicheckers [left] but not for minichess [right]

ADDENDUM

Many readers who experimented with matchbox learning machines were kind enough to write to me about them. L. R. Tanner, at Westminster College, Salt Lake City, Utah, made good use of HER as a concession at a college carnival. The machine was designed to learn by rewards only, so that customers would always have a chance (though a decreasing one) of winning, and prizes to winners were increased in value as HER became more proficient.

Several readers built two matchbox machines to be pitted against each other. John Chambers, Toronto, called his pair THEM (Two-way Hexapawn Educable Machines). Kenneth W. Wiszowaty, science teacher at Phillip Rogers Elementary School, Chicago, sent me a report by his seventh-grade pupil, Andrea Weiland, on her two machines which played against each other until one of them learned to win every time. John House, Waterville, Ohio, called his second machine RAT (Relentless Auto-learning Tyrant), and reported that after eighteen games RAT conceded that HER would win all subsequent games.

Peter J. Sandiford, director of operations research for Trans-Canada Air Lines, Montreal, called his machines Mark I and Mark II. As expected, it took eighteen games for Mark I to learn how to win every time and Mark II to learn how to fight the longest delaying action. Sandiford then devised a devilish plan. He arranged for two students, a boy and a girl from a local high school mathematics club, who knew nothing about the game, to play hexapawn against each other after reading a handout describing the rules. "Each contestant was alone in a room," writes Sandiford, "and indicated his moves to a referee. Unknown to the players the referees reported to a third room containing the jellybean computers and scorekeepers. The players thought they were playing each other by remote control, so to speak, whereas they were in fact playing independently against the computers. They played alternately black and white in successive games. With much confusion and muffled hilarity we in the middle tried to operate the computers, keep the games in phase, and keep the score."

The students were asked to make running comments on their own moves and those of their opponent. Some sample remarks:

"It's the safest thing to do without being captured; it's almost sure to win."

"He took me, but I took him too. If he does what I expect, he'll take my pawn, but in the next move I'll block him."

"Am I stupid!"

"Good move! I think I'm beat."

"I don't think he's really thinking. By now he shouldn't make any more careless mistakes."

"Good game. She's getting wise to my action now."

"Now that he's thinking, there's more competition."

"Very surprising move . . . couldn't he see I'd win if he moved forward?"

"My opponent played well. I guess I just got the knack of it first."

When the students were later brought face to face with the machines they had been playing, they could hardly believe, writes Sandiford, that they had not been competing against a real person.

Richard L. Sites, at M.I.T., wrote a FORTRAN program for an IBM 1620 so that it would learn to play Octapawn, a 4×4 version of hexapawn that begins with four white pawns on the first row and four black pawns on the fourth row. He reports that the first player has a sure win with a corner opening. At the time of his writing, his program had not yet explored center openings.

Judy Gomberg, Maplewood, New Jersey, after playing against a matchbox machine that she built, reported that she learned hexapawn faster than her machine because "every time it lost I took out a candy and ate it."

Robert A. Ellis, at the computing laboratory, Ballistics Research Laboratories, Aberdeen Proving Ground, Maryland, told me about a program he wrote for a digital computer which applied the matchbox-learning technique to a ticktacktoe-learning machine. The machine first plays a stupid game, choosing moves at random, and is easily trounced by human opponents. Then the machine is allowed to play two thousand games against itself (which it does in two or three minutes), learning as it goes. After that, the machine plays an excellent strategy against human opponents.

My defense of Botvinnik's remark that computers will some day play master chess brought a number of irate letters from chess players. One grandmaster assured me that Botvinnik was speaking with tongue in cheek. The interested reader can judge for himself by reading a translation of Botvinnik's speech (which originally appeared in *Komsomolskaya Pravda*, January 3, 1961) in *The Best in Chess*, edited by I. A. Horowitz and Jack Straley Battell (New York: Dutton, 1965), pages 63–69. "The time will come," Botvinnik concludes, "when mechanical chessplayers will be awarded the title of International Grandmaster . . . and it will be necessary to promote two world championships, one for humans, one for robots. The latter tournament, naturally, will not be between machines, but between their makers and program operators."

An excellent science-fiction story about just such a tournament, Fritz Leiber's "The 64-Square Madhouse," appeared in *If*, May 1962, and has since been reprinted in Leiber's *A Pail of Air* (New York: Ballantine, 1964). Lord Dunsany, by the way, has twice given memorable descriptions of chess games played against computers. In his short story "The Three Sailors' Gambit" (in *The Last Book of Wonder*) the machine is a magic crystal. In his novel *The Last Revolution* (a 1951 novel about the computer revolution that has never, unaccountably, been published in the United States) it is a learning computer. The description of the narrator's first game with the computer, in the second chapter, is surely one of the funniest accounts of a chess game ever written.

The hostile reaction of master chess players to the suggestion that computers will some day play master chess is easy to understand; it has been well analyzed by Paul Armer in a Rand report (p-2114–2, June 1962) on *Attitudes Toward Intelligent Machines*. The reaction of chess players is particularly amusing. One can make out a good case against computers writing top-quality music or poetry, or painting great art, but chess is not essentially different from ticktacktoe except in its enormous complexity, and learning to play it well is precisely the sort of thing computers can be expected to do best.

Master checker-playing machines will undoubtedly come first. Checkers is now so thoroughly explored that games between champions almost always end in draws, and in order to add interest to such games, the first three moves are now chosen by chance. Richard Bellman, writing "On the Application of Dynamic Programming to the Determination of Optimal Play in Chess and Checkers," *Proceedings of the National Academy of Sciences,* Vol. 53 (February 1965), pages 244–47, says that "it seems safe to predict that within ten years checkers will be a completely decidable game."

Chess is, of course, of a different order of complexity. One suspects it will be a long time before one can (so goes an old joke in modern dress) play the first move of a chess game against a computer and have the computer print, after a period of furious calculation, "I resign." In 1958 some responsible mathematicians predicted that within ten years computers would be playing master chess, but this proved to be wildly overoptimistic. Tigran Petrosian, when he became world chess champion, was quoted in *The New York Times* (May 24, 1963) as expressing doubts that computers would play master chess within the next fifteen or twenty years.

Hexapawn can be extended simply by making the board wider but keeping it three rows deep. John R. Brown, in his paper "Extendapawn—An Inductive Analysis," *Mathematics Magazine,* Vol. 38, November 1965, pages 286–99, gives a complete analysis of this game. If n is the number of columns, the game is a win for the first player if the final digit of n is 1, 4, 5, 7 or 8. Otherwise the second player has the win.

ANSWERS

The checker game on the 4×4 board is a draw if both sides play as well as possible. As shown in Figure 47, Black has a choice of three openings: (1) C5, (2) C6, (3) D6.

The first opening results in an immediate loss of the game when White replies A3. The second opening leads to a draw regardless of how White replies. The third opening is Black's strongest. It leads to a win if White replies A3 or B3. But White can reply B4 and draw.

With respect to the 3×3 simplified go game, also mentioned as suitable for a matchbox learning machine, I am assured by Jay Eliasberg, vice-president of the American Go Association, that the first player has a sure win if he plays on the center point of the board and rationally thereafter.

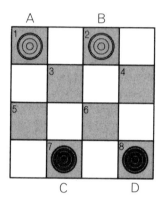

Figure 47
Checker game is drawn if
played rationally

The 4 × 4 checker game is trivial, but when the board is enlarged to 5 × 5 the result is both challenging and surprising. Robert L. Caswell, a chemist with the United States Department of Agriculture, wrote to me about this mini-game, which he said had earlier been proposed to him. The game begins with three white checkers on the first row, three black checkers on the fifth row. All standard rules obtain, with black moving first. One might guess the game to be drawn if played rationally, but the absence of "double corners" where kings can move back and forth makes this unlikely. Caswell discovered that not only does one side have a sure win but, if the loser plays well, the final win is spectacular. Rather than spoil the fun, I leave it to the reader to analyze the game and decide which player can always win.

CHAPTER NINE

O

Spirals

The spiral is a spiritualized circle. In the spiral form, the circle, uncoiled, unwound, has ceased to be vicious; it has been set free. I thought this up when I was a schoolboy, and I also discovered that Hegel's triadic series expressed merely the essential spirality of all things in their relation to time. Twirl follows twirl, and every synthesis is the thesis of the next series. . . . A colored spiral in a small ball of glass, this is how I see my own life.
—VLADIMIR NABOKOV, *Speak Memory*

TWO FARM CHILDREN have improvised a seesaw by placing a plank over a log. As they go up and down, what sort of curve is traced by every point along the plank? On a moving carrousel the operator walks at a constant speed along a radius of the floor. What type of curve does he trace on the ground beneath the carrousel? Three dogs stand in an open field at the corners of an equilateral triangle. On command each dog runs directly toward the dog on its right. Turning to follow one another as they move, all three run with the same constant speed until they meet at the triangle's center. What sort of paths do they take?

The answer to each question is a different type of spiral. I shall describe the three curves in turn and in doing so try to spiral around as many recreational sidelights as space allows.

The curves traced by all points along the plank of the seesaw are known as involutes of the circle. The involute of any curve is obtained by attaching a thread to the curve, pulling it taut, then "winding" it along the curve. Any fixed point on the taut thread traces the curve's involute. Thus a goat tied to a cylindrical post will, if it circles the post so that the rope winds tightly around it, be pulled into a spiral path that is the involute of a circle.

A neat way to draw such a spiral is depicted in Figure 48. Cut a circle of any desired size from thick cardboard and cement it to the center of a sheet of paper. Cement a slightly larger circle of cardboard on top, with a slot on the rim to hold the knotted end of a piece of string. Wind the string around the smaller circle. The point of a pencil, in a loop at the free end of the cord, will unwind the string and trace the involute. The distance between adjacent coils remains constant and is equal to the smaller circle's circumference when measured along a line that is tangent to one side of the circle. The circle is said to be the evolute of the spiral.

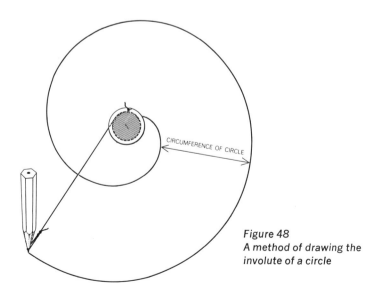

CIRCUMFERENCE OF CIRCLE

Figure 48
A method of drawing the
involute of a circle

The man on the carrousel traces (with respect to the ground) a curve known as the spiral of Archimedes. (Archimedes was the first to study it; his treatise *On Spirals* is concerned mainly with this curve.) If you place a cardboard disk on a phonograph turntable, you can draw on it a spiral of Archimedes by moving a crayon at a constant speed in a straight line from the center of the disk outward. The groove in a phonograph record is the most familiar example of such a spiral. In polar coordinates it is described by saying that at every point the radius vector (distance from the disk's center) is in the same ratio to the vector angle (angular distance from a fixed radius). Spirals have very simple equations in polar coordinates but very complicated equations in Cartesian coordinates.

A much more accurate Archimedean spiral can be obtained by pinning a strip of cardboard, cut as shown in Figure 49,

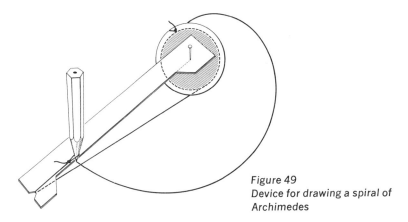

Figure 49
Device for drawing a spiral of Archimedes

to a pair of cardboard circles like those used for drawing the involute. As the strip is revolved, the pencil point will be pulled outward along one edge of the strip. It is easy to see that the pencil must move along the edge with a speed that is always proportional to the speed at which the cardboard strip is revolving.

After the first turn the resulting spiral is virtually indistinguishable from the involute of a circle, although the two curves are never exactly alike. The distance between adjacent coils of the Archimedean spiral is constant, but now the distance must be measured along radii instead of along lines tangent to one side of a circle. The most commonly observed spirals are of the Archimedean or circle-involute types: tightly wound springs, edges of rolled-up rugs and sheets of paper, decorative spirals on jewelry, and so on. Such curves are seldom mathematically precise, and one would be hard put to determine whether a given example is in fact closer to a circle involute or a spiral of Archimedes.

Once an accurate Archimedean spiral has been drawn, it can be used for compass-and-straightedge divisions of any angle into any number of equal parts, including three. To trisect an angle, place the angle so that its vertex coincides with the spiral's pole (origin) and its arms intersect the spiral (*see Figure 50*). With the point of the compass at P, draw arc AB. The line segment AC is trisected by the usual

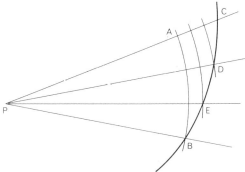

Figure 50
Trisecting an angle with a spiral of Archimedes

method. Through the two points between A and C thus established, arcs of circles are drawn to mark points D and E on the spiral. Lines from the vertex to D and E complete the trisection. Readers may enjoy proving that this construction is accurate.

The mechanical device pictured in Figure 51 is often used in machines for transforming the uniform circular motion of a wheel into a uniform back-and-forth motion. (Many sewing machines, for example, use such a device for moving the

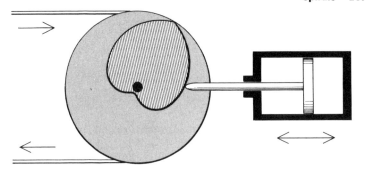

Figure 51
Archimedean spirals change rotary to linear motion

thread back and forth when the bobbins are wound.) The sides of the heart are mirror-image arcs of a spiral of Archimedes.

The dogs that chase one another to the center of the equilateral triangle follow the lines of a logarithmic, or equiangular, spiral. One way to define this spiral is to say that it cuts every radius vector at the same angle. If mathematical points are substituted for dogs, each point traces a path of finite length (it is two thirds the side of the triangle), but only after making an infinite number of revolutions around the pole! Logarithmic spirals also mark the paths of any number of dogs greater than two, provided that they start at the corners of a regular polygon. If there are only two dogs, their paths are, of course, straight lines; if there are an infinite number, they keep trotting around a circle. This is a crude way of pointing out that the limits of the equiangular spiral, as its angle to the radius vector varies from 0 to 90 degrees, are the straight line and the circle.

On the earth's surface the counterpart of the logarithmic spiral is the loxodrome (or rhumb line) : a path that cuts the earth's meridians at any constant angle except a right angle. Thus if you were flying northeast and always kept the plane heading in exactly the same direction as indicated by the compass, you would follow a loxodrome that would spiral you to the North Pole. Like the dogs' paths, your path to the Pole would be finite in length but (if you were a point) you would have to circle the Pole an infinite number of times before you got there. A stereographic projection of your path on a plane tangent to the Pole would be a perfect logarithmic spiral.

The logarithmic spiral is the most common type of spiral to be found in nature. It can be seen in the coil of the nautilus shell and snail shells, in the arrangement of the seeds of many plants, such as the sunflower and daisy, the scales of the pine cone, and so on. *Epeira*, a common variety of spider, spins a web in which a strand coils around the center in a logarithmic spiral. Jean Henri Fabre, in his book *The Life of the Spider*, devotes an appendix to a discussion of the mathematical properties of the equiangular spiral and its many beautiful appearances in nature. There is an extensive literature, some of it eccentric, on this spiral's botanical and zoological manifestations and its close relation to the golden ratio and the Fibonacci number series. The basic reference here is a 479-page, richly illustrated book entitled *The Curves of Life*, by Theodore Andrea Cook. It was published in 1914 by Henry Holt and has long been out of print.

A device for ruling a logarithmic spiral is easily cut from a piece of cardboard (*see Figure 52*). Angle *a* may be any size you please between 0 and 180 degrees. By keeping one edge of the strip on the spiral's pole and ruling short line segments along the oblique straightedge as this straightedge is moved toward or away from the pole, you produce a series

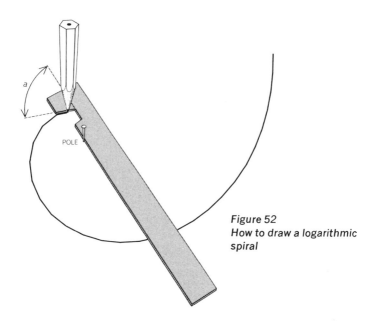

Figure 52
How to draw a logarithmic spiral

of chords of the spiral in much the same manner that *Epeira* spins its web. The device ensures that all these chords cut the radius vector at the same angle. The smaller you make the oblique straightedge, of course, the more accurate the spiral is. Such a device can also be used for testing a spiral to see if it is logarithmic.

What happens if angle *a* is a right angle? The spiral degenerates into a circle. If the angle is 74 degrees 39 minutes (the exact value is a trifle more than this), the resulting spiral will be its own involute. The involutes of all logarithmic spirals are also logarithmic spirals, but only in this case are the two spirals exactly alike.

The equiangular spiral was first discovered by René Descartes. Jakob Bernoulli, the seventeenth-century Swiss mathematician, was so entranced by the spiral's property of reappearing after various transformations (e.g., changing it to its involute) that he asked to have it engraved on his tombstone with the words *"Eadem mutata resurgo"* ("Though changed I shall arise the same"). His request was badly carried out. The Latin phrase was omitted, and the best spiral the poor stonecutter could achieve was a crude version of either an Archimedean spiral or an involute of a circle. It can be seen today on the mathematician's gravestone in Basel, and it is obviously not a logarithmic spiral, because the width between coils shows no progressive increase as it grows larger.

In terms of sheer size, the logarithmic spiral's most impressive appearance is in the arms of many of the spiral galaxies. Just why it turns up here is a mystery that is bound up with the mystery of the arms themselves. They are known to be glowing lanes of stars and gas that somehow are whirled into spiral shape by the galaxy's rotation. The entire galaxy is a cluster of billions of stars and spins like a monstrous Fourth of July pinwheel. The faint white glow of the Milky Way results from our looking edgewise through two gigantic spiral arms of our own galaxy. Observations show that these arms are rotating much faster near the center of the galaxy than at the edge. This ought to wind up the arms quickly and eventually eliminate them, but the fact that most galaxies have retained a spiral structure suggests that the arms are not winding up at all. One theory has it that as one side of an arm takes on luminous gas, the other side evaporates it, keeping the arm in the same shape with respect to the galaxy (see Jan H. Oort, "The Evolution of Galaxies," *Scientific American*, September 1956).

Like their space-curve cousin the helix, all spiral shapes are asymmetric. This means that on a plane every spiral can be drawn in two forms that are identical in all respects except that one is a mirror reflection of the other. When a spiral can be viewed from either side, as is the case with spider webs and (if we could travel far enough out in space) galaxies, then its "handedness" depends on the point of view. But if there is no way to turn a spiral over or to move around in order to see it from the other side, every spiral is either clockwise or counterclockwise.

The adjective "clockwise" is ambiguous, of course, unless you specify whether the spiral is traced outward from the center or inward toward the center. There is an amusing pencil-and-paper stunt based on this ambiguity. Ask someone to draw a spiral on the left side of a sheet, starting at the center and moving the pencil outward. Cover the spiral with your hand and ask him to draw a mirror-image of that spiral on the right of the sheet, starting with a large loop and spiraling in to the center. Most people will reverse the rotary motion of their hand, but of course this simply produces another spiral of the same handedness.

If you draw a tightly coiled spiral on a cardboard disk, using a thick black line, and rotate it on a phonograph turntable, a familiar illusion results. The coils appear either to expand or contract depending on the spiral's handedness. An even more astonishing psychological illusion can be demonstrated with two such disks, bearing spirals of opposite handedness. Put the "expanding" spiral on the turntable and stare directly down at its pole for several minutes while it revolves. Now quickly shift your gaze to someone's face. For a moment the face will appear to shrink suddenly. The other spiral has the opposite effect: the face you look at will appear to explode outward. Everyone has experienced a similar illusion when riding on a train. After looking for a long time out the window of a moving train, if the train stops, the scenery momentarily seems to move in the opposite direction. There have been attempts to explain this in terms of eye movements and the tiring of eye muscles, but the spiral illusion rules out such an explanation. It suggests that the illusion arises in the brain's interpretation of signals from the eyes.

The asymmetry of the spiral makes it a convenient figure for dramatizing a curious problem of communication. Imagine that Project Ozma has established radio-wave contact with a Planet X somewhere in our galaxy. Over the decades, by the use of ingenious pulsed codes, we learn to converse fluently with intelligent humanoids on Planet X. It has a culture almost as advanced as ours but because of high, dense clouds like the clouds of Venus surrounding it, its inhabitants know nothing about astronomy. They have never seen the stars. After Planet X has been sent a detailed description of a number of major galaxies, the following message is received on earth:

"You say spiral nebula NGC 5194, viewed from earth, has two spiral arms that coil outward in a clockwise direction. Please clarify meaning of 'clockwise.' "

In other words, scientists on Planet X want to be sure that when they record a diagram of nebula NGC 5194, based on information supplied by scientists on earth, they draw it correctly and not in mirror-image form.

How can we communicate to Planet X which way the nebula coils? It is no help to say that as an arm whirls outward above the center of the galaxy it moves from left to right, because we have no way of being certain that Planet X understands "left" and "right" in the same way we do. If we could communicate an unambiguous definition of "left," the problem would of course be solved.

To give the problem more precisely: How can we communicate the meaning of "left" by a language transmitted as a pulsed code? We may say anything we please, request the performance of any type of experiment, with one proviso: There is to be no asymmetric object or structure that we and they can observe in common.

Without this proviso there is no problem. For example, if we sent to Planet X a rocket missile carrying a picture of a man labeled "top," "bottom," "left," "right," the picture would immediately convey our meaning of "left." Or, we might transmit a radio beam that had been given a helical twist by circular polarization. If the inhabitants of Planet X built antennas that could determine whether the polarization was clockwise or counterclockwise, a common understanding of "left" could easily be established. Such methods, however, violate the proviso that there must be no common observation of a particular asymmetric object or structure.

ANSWERS

It is easy to see why the trisection of an angle, by means of an Archimedean spiral, works. The arcs of the circles mark, along the spiral, three equal distances along the radius vector; that is, equal distances away from P, the vertex of the angle to be trisected. As the spiral travels those distances outward, it also travels equal distances in a counterclockwise direction, creating three equal vectorial angles. The same method obviously can be used for dividing an angle into any number of smaller angles in any desired ratio to each other. Simply divide line segment AC into segments with the desired ratios. The construction will then divide angle CPB into those same ratios.

How can the meaning of our word "clockwise" be communicated by a pulsed code to humanoids on Planet X? It is assumed that Planet X is somewhere in our galaxy but covered by dense clouds that prevent its inhabitants from seeing the stars. It is also assumed that by ingenious codes scientists on earth and on Planet X have learned to talk fluently with each other. The problem is how to communicate the meaning of "left" and "right."

The startling answer is that until December 1956 there was *no* way to communicate an unambiguous definition of "left" and "right." According to what physicists call the "law of parity," all asymmetric physical processes are reversible; that is, they can take place in either of their two mirror-image forms. Certain crystals, such as quartz and cinnabar, have the property of twisting a plane of polarized light in one direction only, but such crystals exist in both left and right forms. The same is true of the asymmetric stereoisomers, which also twist planes of polarized light. Organic compounds found in living forms may possess one type of handedness only, but this is an accident of the earth's evolution. There is no more reason for such compounds on another planet having the same handedness as those on earth as there would be reason to expect the humanoids on Planet X to have hearts on their left sides.

Electrical and magnetic experiments are of no help. It is true that they show asymmetries (e.g., the "right-hand rule" for orienting a magnetic field surrounding a current), but it is only convention that decides which pole of a magnet is called "north." If we could communicate to Planet X what we mean by a "North Pole," the problem could be solved; unfortunately there is no way to do this without first having a common understanding of *left* and *right*. We could easily transmit pictures to Planet X by means of pulsed codes, but without agreement on left and right we could never be sure that their equipment was not reproducing the pictures in a form that was the reverse of ours.

It was in December 1956 that the first experiment violating the law of parity was performed (see Philip Morrison, "The Overthrow of Parity," *Scientific American*, April 1957). Certain "weak interactions" of particle physics were found to show a preference for one type of handedness regardless of the North Pole–South Pole convention. Sending the details of such an experiment is the only way known at present by which we could communicate to Planet X an unambiguous operational definition of left and right, clockwise and counterclockwise, the North and the South Magnetic Pole, or any other distinction involving handedness.

It should be added that if Planet X were in another galaxy, the problem would remain unsolved. The other galaxy might be made of antimatter (matter made of particles with reversed electrical charges). In such a galaxy the handedness of the weak interactions would probably be reversed. If we did not know the type of matter in the other galaxy (and light from it provides no clue), parity-violating experiments would be valueless in communicating the meaning of *left* and *right*.

CHAPTER TEN

O

Rotations
and Reflections

A GEOMETRIC FIGURE is said to be symmetrical if it remains unchanged after a "symmetry operation" has been performed on it. The larger the number of such operations, the richer the symmetry. For example, the capital letter A is unchanged when reflected in a mirror placed vertically beside it. It is said to have vertical symmetry. The capital B lacks this symmetry but has horizontal symmetry: it is unchanged in a mirror held horizontally above or below it. S is neither horizontally nor vertically symmetrical but remains the same if rotated 180 degrees (twofold symmetry). All three of these symmetries are possessed by H, I, O and X. X is richer in symmetry than H or I because, if its arms cross at right angles, it is also unchanged by quarter-turns (fourfold symmetry). O, in circular form, is the richest letter of all. It is unchanged by any type of rotation or reflection.

Because the earth is a sphere toward the center of which all objects are drawn by gravity, living forms have found it efficient to evolve shapes that possess strong *vertical* symmetry combined with an obvious lack of horizontal or rotational symmetry. In making objects for his use man has followed a similar pattern. Look around and you will be struck by the number of things you see that are essentially unchanged in a vertical mirror: chairs, tables, lamps, dishes, automobiles, airplanes, office buildings—the list is endless. It is this prevalence of vertical symmetry that makes it so difficult to tell when a photograph has been reversed, unless the scene is familiar or contains such obvious clues as reversed printing or cars driving on the wrong side of the road. On the other hand, an upside-down photograph of almost anything is instantly recognizable as inverted.

The same is true of works of graphic art. They lose little, if anything, by reflection, but unless they are completely non-representational no careless museum director is likely to hang one upside down. Of course, abstract paintings are often inverted by accident. *The New York Times Magazine* (October 5, 1958) inadvertently both reversed and inverted a picture of an abstraction by Piet Mondrian, but only readers who knew the painting could possibly have noticed it. In 1961, at the New York Museum of Modern Art, Matisse's painting, *Le Bateau,* hung upside down for forty-seven days before anyone noticed the error.

So accustomed are we to vertical symmetry, so unaccustomed to seeing things upside down, that it is extremely difficult to imagine what most scenes, pictures or objects would look like inverted. Landscape artists have been known to check the colors of a scene by the undignified technique of bending over and viewing the landscape through their legs. Its upside-down contours are so unfamiliar that colors can be seen uncontaminated, so to speak, by association with familiar shapes. Thoreau liked to view scenes this way and refers to such a view of a pond in Chapter 9 of *Walden.* Many philosophers and writers have found symbolic meaning in this vision of a topsy-turvy landscape; it was one of the favorite themes of G. K. Chesterton. His best mystery stories (in my opinion) concern the poet-artist Gabriel Gale (in *The Poet and the Lunatics*), who periodically stands on his hands so that he can "see the landscape as it really is: with the stars like flowers, and the clouds like hills, and all men hanging on the mercy of God."

The mind's inability to imagine things upside down is essential to the surprise produced by those ingenious pictures that turn into something entirely different when rotated 180 degrees. Nineteenth-century political cartoonists were fond of this device. When a reader inverted a drawing of a famous public figure, he would see a pig or jackass or something equally insulting. The device is less popular today, although *Life* for September 18, 1950, reproduced a remarkable Italian poster on which the face of Garibaldi became the face of Stalin when viewed upside down. Children's magazines sometimes reproduce such upside-down pictures, and now and then they are used as advertising gimmicks. The back cover of *Life* for November 23, 1953, depicted an Indian brave inspecting a stalk of corn. Thousands of readers probably failed to notice that when this picture was inverted it became the face of a man, his mouth watering at the sight of an open can of corn.

I know of only four books that are collections of upside-down drawings. Peter Newell, a popular illustrator of children's books who died in 1924, published two books of color plates of scenes that undergo amusing transformations when inverted: *Topsys & Turvys* (1893) and *Topsys and Turvys Number 2* (1894). In 1946 a London publisher issued a collection of fifteen astonishing upside-down faces drawn by Rex Whistler, an English muralist who died in 1944. The book has the richly symmetrical title of *¡OHO!* (Its title page is reproduced in Figure 53.)

Figure 53
Invertible faces on the title page of Whistler's invertible book

The technique of upside-down drawing was carried to un-believable heights in 1903 and 1904 by a cartoonist named Gustave Verbeek. Each week he drew a six-panel color comic for the Sunday "funny paper" of the *New York Herald*. One took the panels in order, reading the captions beneath each picture; then one turned the page upside down and continued the story, reading a new set of captions and taking the same six panels in reverse order! (*See Figure 54*.) Verbeek man-aged to achieve continuity by means of two chief characters

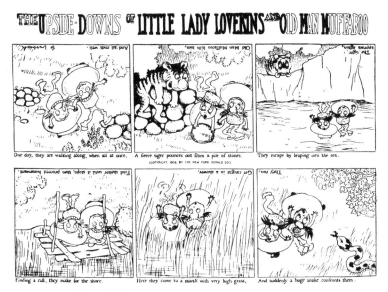

Figure 54
A typical upside-down cartoon by Gustave Verbeek

called Little Lady Lovekins and Old Man Muffaroo. Each be-came the other when inverted. How Verbeek managed to work all this out week after week without going mad passeth all understanding. A collection of twenty-five of his comics was published by G. W. Dillingham in 1905 under the title of *The Upside-Downs of Little Lady Lovekins and Old Man Muffaroo*. The book is extremely rare.

The 90-degree rotation is less frequently used in art play, perhaps because it is easier for the mind to anticipate re-sults. If done artfully, however, it can be effective. An exam-ple is a landscape by the seventeenth-century Swiss painter

Matthäus Merian that becomes a man's profile when the picture is given a quarter-turn counterclockwise. The rabbit-duck in Figure 55 is the best-known example of a quarter-turn picture. Psychologists have long used it for various sorts of testing. A few years ago Harvard philosopher Morton White reproduced a rabbit-duck drawing in a magazine article to symbolize the fact that two historians can survey the same set of historical facts but see them in two essentially different ways.

Our lifelong conditioning in the way we see things is responsible for a variety of startling upside-down optical illusions. All astronomers know the necessity of viewing photographs of the moon's surface so that sunlight appears to illuminate the craters from above rather than below. We are so unaccustomed to seeing things illuminated from below that when such a photograph of the moon in inverted, the craters instantly appear to be circular mesas rising above the surface. One of the most amusing illusions of this same general type is shown in Figure 56. The missing slice of pie is found by turning the picture upside down. Here again the explanation surely lies in the fact that we almost always see plates and pies from above and almost never from below.

Figure 55
A quarter-turn clockwise makes
the duck into a rabbit

Figure 56
Where is the missing slice?

Upside-down faces could not be designed, of course, if it were not for the fact that our eyes are not too far from midway between the top of the head and the chin. School children often amuse themselves by turning a history book upside down and penciling a nose and mouth on the forehead of some famous person.

When this is done on an actual face, using eyebrow pencil

and lipstick, the effect becomes even more grotesque. It was a popular party pastime of the late nineteenth century. The following account is from an old book entitled *What Shall We Do Tonight?*

> The severed head always causes a sensation and should not be suddenly exposed to the nervous. . . . A large table, covered with a cloth sufficiently long to reach to the floor all around and completely hide all beneath, is placed in the center of the room. . . . A boy with soft silky hair, rather long, being selected to represent the *head*, must lie upon his back under the table entirely concealed, excepting that portion of his face above the bridge of his nose. The rest is under the tablecloth.
>
> His hair must now be carefully combed down, to represent whiskers, and a face must be painted . . . upon the cheeks and forehead; the false eyebrows, nose and mouth, with mustache, must be strongly marked with black water color, or India ink, and the real eyebrows covered with a little powder or flour. The face should also be powdered to a deathlike pallor. . . .
>
> The horror of this illusion may be intensified by having a subdued light in the room in which the exhibition has been arranged. This conceals in a great degree any slight defects in the "making-up" of the head. . . .

Needless to add, the horror is heightened when the "head" suddenly opens its eyes, blinks, stares from side to side, wrinkles its cheeks (forehead).

The physicist Robert W. Wood (author of *How to Tell the Birds from the Flowers*) invented a funny variation of the severed head. The face is viewed upside down as before, but now it is the forehead, eyes and nose that are covered, leaving only the mouth and chin exposed. Eyes and nose are drawn on the chin to produce a weird little pinheaded creature with a huge, flexible mouth. The stunt is a favorite of Paul Winchell, the television ventriloquist. He wears a small dummy's body on his head to make a figure that he calls Ozwald, while television camera techniques invert the screen to bring Ozwald right side up. In 1961 an Ozwald kit was marketed for children, complete with the dummy's body and a special mirror with which to view one's own face upside down.

It is possible to print or even write in longhand certain words in such a way that they possess twofold symmetry. The Zoological Society of San Diego, for instance, publishes a magazine called *ZOONOOZ*, the name of which is the same upside down. The longest sentence of this type that I have come across is said to be a sign by a swimming pool designed to read the same when viewed by athletes practicing handstands: NOW NO SWIMS ON MON. (*See Figure 57.*)

Figure 57
An invertible sign [sketch reproduced by courtesy of the artist, John McClellan]

It is easy to form numbers that are the same upside down. As many have noticed, 1961 is such a number. It was the first year with twofold symmetry since 1881, the last until 6009, and the twenty-third since the year 1. Altogether there are thirty-eight such years between A.D. 1 and A.D. 10000 (according to a calculation made by John Pomeroy), with the longest interval between 1961 and 6009. J. F. Bowers, writing in the *Mathematical Gazette* for December 1961, explains his clever method of calculating that by A.D. 1000000 exactly 198 invertible years will have passed. The January 1961 issue of *Mad* featured an upside-down cover with the year's numerals in the center and a line predicting that the year would be a mad one.

Some numbers, for example 7734 (when the 4 is written so that it is open at the top), become words when inverted; others can be written to become words when reflected. With these quaint possibilities in mind, the reader may enjoy tackling the following easy problems:

1. Oliver Lee, age forty-four, who lives at 312 Main Street, asked the city to give his car a license plate bearing the number 337-31770. Why?

2. Prove the sum in Figure 58 to be correct.

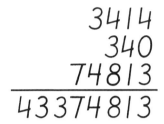

Figure 58
Is the sum correct?

3. Circle six digits in the group below that will add up to exactly 21.

1	1	1
3	3	3
5	5	5
9	9	9

4. A basket contains more than half a dozen eggs. Each egg is either white or brown. Let x be the number of white eggs, y the number of brown. The sum of x and y, turned upside down, is the product of x and y. How many eggs are in the basket?

ANSWERS

1. The number 337-31770 upside down spells "Ollie Lee."

2. Hold the sum to a mirror.

3. Turn the picture upside down, circle three 6's and three 1's to make a total of 21.

4. The basket has nine white eggs and nine brown eggs. When the sum, 18, is inverted, it becomes 81, the product. Had it not been specified that the basket contained more than six eggs, three white and three brown would have been another answer.

CHAPTER ELEVEN

O

Peg Solitaire

"THE GAME CALLED SOLITAIRE pleases me much," the great German mathematician Gottfried von Leibniz wrote in a letter in 1716. "I take it in reverse order. That is to say, instead of making a figure according to the rules of the game, which is to jump to an empty place and remove the piece over which one has jumped, I thought it better to reconstruct what had been demolished by filling an empty hole over which one has leaped. In this way one may set oneself the task of forming a given figure if that is possible, as it certainly is if it can be destroyed. But why all this? you ask. I reply: to perfect the art of invention. For we must have the means of constructing everything which is found by the exercise of reason."

Leibniz's last two sentences are a bit obscure. Perhaps they mean that it is worthwhile to analyze everything that has a logical or mathematical structure.

Worthwhile or not, no other puzzle game played on a board with counters has enjoyed such a long, uninterrupted run of popularity as solitaire. Its origin is unknown, although its invention is sometimes attributed to a prisoner in the Bastille. That it was widely played in France during the late nineteenth century is evident from the many French books and articles that were then written about the game. It is likely that almost every reader of this column has at one time or another racked his brain over the puzzle. At present several versions of solitaire are on sale in this country under various trade names, some with pegs that are moved from hole to hole and some with marbles that rest in circular depressions. The marble versions are easier to manipulate. One can also play by placing pennies, beans, small poker chips or any other type of counter on the board depicted in Figure 59.

This board, which has thirty-three cells, is the most popular form of solitaire in England, the United States and the U.S.S.R. In France the board has four additional cells at the positions indicated by the four dots. Both forms of the board are found throughout the rest of Western Europe, but the French form has been the least popular, probably because it is not possible to reduce the full board, with center vacant, to a single peg. The cells are labeled in traditional fashion, the first digit of each number giving the position of the column from left to right, the second digit giving the position of the row from bottom to top.

The basic problem—usually the only problem supplied by manufacturers of the puzzle—begins with counters placed on all cells except the center one. The object is to make a series of jumps that will remove every counter but one. For an elegant solution this last counter should be left on the central cell. A "jump" consists of moving a counter over any adjacent counter to land on the next empty cell. The jumped counter is taken off the board. This is the same as a jump in checkers except that each jump must be straight to the left

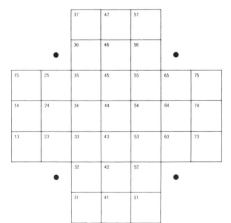

Figure 59
The solitaire board

or right, or straight up or down. No diagonal jumps are allowed.

Each move must be a jump. If a point is reached at which no jumps are possible, the game ends in a stalemate. A single piece may continue in a chain of connected jumps as long as jumps are available, but it need not do so. A chain of jumps is counted as a single "move." To solve the puzzle, thirty-one jumps, obviously, must be made, but if some are in chains, the number of moves can be fewer.

No one knows how many different ways there are to solve the puzzle leaving the last counter in the center. Scores of solutions have been published. Before discussing some of them, however, readers unfamiliar with solitaire are urged to try the six simpler figures shown in Figure 60. In each case the last counter must be left on the center cell. For example, the Latin cross is easily solved in five moves: 45-25, 43-45, 55-35, 25-45, 46-44.

After mastering these traditional problems the reader may want to try the three puzzles shown in Figure 61. In each of these one must begin with a full board, except for a vacant center cell, and play until the figure shown remains on the board. The first puzzle is easy; the other two are not. Note that the pinwheel is a stalemated position. It is possible to reach a stalemate in as few as six moves. Can you discover how?

Advanced students of solitaire have gone to fantastic lengths in setting themselves unusual tasks. For example, in his book *The Game of Solitaire* (1920) Ernest Bergholt introduces into his brilliant problems a variety of curious restrictions. (All the problems start with a full board, although the vacant cell need not be in the center.) His "ball on the watch" is a single counter—preferably a different color from the others—that must not be moved until the end of the game; then it captures one or more pieces to become the sole survivor. His "dead ball" is a counter that remains untouched throughout and is the last to be taken. A "sweep" is a long chain of jumps that closes a game. Bergholt gives many examples of games ending in eight-ball sweeps. It is possible, he maintains, to begin with the vacancy at a corner cell, say 37, and end with a nine-ball sweep.

What is the smallest number of moves required to reduce

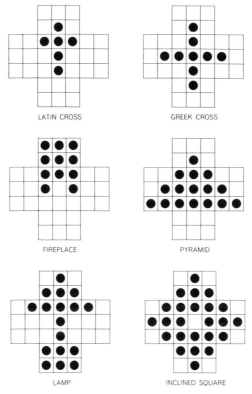

LATIN CROSS

GREEK CROSS

FIREPLACE

PYRAMID

LAMP

INCLINED SQUARE

Figure 60
Traditional problems in which the last counter
is to be left in the center

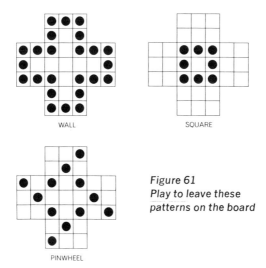

WALL

SQUARE

PINWHEEL

Figure 61
Play to leave these
patterns on the board

a full board of thirty-two pieces to a single piece? It had long been thought that sixteen moves was minimal, but in 1963 Harry O. Davis, of Portland, Oregon, found fifteen-move solutions when the initial vacancy is cell 55 or 52 and the cells that correspond to those two when the board is rotated and reflected. Here is Davis's solution with the vacancy at 55 and the last counter also on 55: 57-55, 54-56, 52-54, 73-53, 43-63, 37-57-55-53, 35-55, 15-35, 23-43-45-25, 13-15-35, 31-33, 36-56-54-52-32, 75-73-53, 65-63-43-23-25-45, 51-31-33-35-55. If the vacancy is at 52, the other fifteen-move solution discovered by Davis ends with the counter at 55.

Davis found sixteen-move solutions when the initial vacancy is 54 or 57 and all symmetrically corresponding cells, and seventeen-move solutions on all other cells (the center, 46, 47, and symmetrically corresponding cells).

There are twenty-one distinct combinations of initial and terminal cells (not counting rotations and reflections, of course). The followng chart lists the minimum-move solutions that have been found by Davis:

Vacancy	Terminal cell	Number of moves
13	13	16
13	43	16
13	46	17
13	73	16
14	14	18
14	41	17
14	44	18
14	74	18
23	23	16
23	53	15
23	56	16
24	24	19
24	51	17
24	54	17
33	33	15
33	63	16
34	31	16
34	34	16
34	64	17
44	14	17
44	44	18

As the chart shows, if the game opens with a vacant center cell (44) and ends with a counter on the same cell, eighteen moves are required. Henry Ernest Dudeney, in his *Amusements in Mathematics* (Problem No. 227), gives a nineteen-move solution and adds: "I do not think the number of moves can be reduced." But Bergholt gives in his book the following eighteen-move solution: 46-44, 65-45, 57-55, 54-56, 52-54, 73-53, 43-63, 75-73-53, 35-55, 15-35, 23-43-63-65-45-25, 37-57-55-53, 31-33, 34-32, 51-31-33, 13-15-35, 36-34-32-52-54-34, 24-44. (Dudeney's solution first appeared in *The Strand Magazine*, April 1908. Bergholt first published his shorter solution in *The Queen*, May 11, 1912.)

"I will venture to assert," writes Bergholt, "that this record will never be beaten." (That eighteen is indeed minimal has recently been proved by J. D. Beasley, at Cambridge University.) Note that if the chain of jumps in Bergholt's next-to-last move is not interrupted, a seventeen-move solution is achieved, ending on cell 14, with the counter originally placed on cell 36 serving as a ball on the watch that closes the game with a six-ball sweep.

Other solutions of the classic center-to-center problem, although failing to achieve the minimum in moves, often have a remarkable symmetry. Consider the following examples.

"The Fireplace" (discovered by James Dow, of Boston): 42-44, 23-43, 35-33, 43-23, 63-43, 55-53, 43-63, 51-53, 14-34-54-52, 31-51-53, 74-54-52, 13-33, 73-53, 32-34, 52-54, 15-35, 75-55. The counters now form the fireplace shown in Figure 60 and the game is completed according to the solution of that problem. This shortens by three moves a similar fireplace solution by Josephine G. Richardson, also of Boston, that is given in *Puzzle Craft*, a booklet edited by Lynn Rohrbough and published in 1930 by the Co-operative Recreation Service of Delaware, Ohio. The next two solutions are from Rohrbough's booklet.

"The Six-Jump Chain": 46-44, 65-45, 57-55, 37-57, 54-56, 57-55, 52-54, 73-53, 75-73, 43-63, 73-53, 23-43, 31-33, 51-31, 34-32, 31-33, 36-34, 15-35, 13-15, 45-25, 15-35. The pattern now has vertical symmetry. A six-jump chain (43-63-65-45-25-23-43) reduces the pattern to a T figure, easily solved with 44-64, 42-44, 34-54, 64-44.

"The Jabberwocky": 46-44, 65-45, 57-55, 45-65, 25-45, 44-

46, 47-45, 37-35, 45-25. The pattern is vertically symmetrical. The next sixteen moves are mirror-image pairs that can be made simultaneously by the right and left hands, as follows:

Left hand	Right hand
15-35	75-55
34-36	54-56
14-34	74-54
33-35	53-55
36-34	56-54
31-33	51-53
34-32	54-52
13-33	73-53

The solution concludes: 43-63, 33-31-51-53, 63-43, 42-44.

The mathematical theory behind solitaire is only partly known. In fact, one of the major unsolved problems of recreational mathematics is finding a way to analyze a given solitaire position to determine whether or not it is possible to reduce it to another given position. A man who has made considerable progress in this direction is Mannis Charosh, a teacher of mathematics at New Utrecht High School in Brooklyn, New York. In *The Mathematics Student Journal* for March 1962, he proves a variety of unusual theorems that combine to provide an extremely useful technique for establishing the impossibility of certain solitaire problems. Charosh's analysis simplifies and extends an earlier analysis by M. H. Hermary, to be found in the first volume of *Récréations Mathématiques,* edited by the French mathematician Édouard Lucas.

Charosh's method consists of applying a series of transformations to any starting position to see if it can be changed to the desired end position. If it can, the two positions are said to be "equivalent." If two positions are *not* equivalent, it is impossible to change one to the other by jumping pegs (or, alternatively, by working backward as Leibniz suggested). If two positions *are* equivalent, the problem may or may not be solvable by the rules of solitaire. In other words, the method gives to any solitaire problem, on any type of board, a necessary but not a sufficient condition of possibility.

Charosh's transformations involve any set of three adjacent cells that are in a straight horizontal or vertical line. Where there are counters on these three cells, remove them; where there are vacancies, put counters. Thus if all three cells are filled, all three counters can be removed. If all three are vacant, all three can be filled. If there are two counters, the two can be removed and a single counter can be placed on the previously empty cell. If there is only one counter, it can be removed and counters can be placed on the two previously empty cells.

Let us apply this method to the classic problem that begins with a vacancy in the center. It can be seen at once that sets of three counters in a row can be removed until only two counters remain on, say, cells 45 and 43. Since these are the ends of the triplet 43, 44, 45, we can remove the two counters and substitute a counter on 44. We have thereby shown that the full board, with an empty cell at 44, is equivalent to an empty board with a single counter on 44; therefore the problem is not impossible. (We already know, of course, that it can be solved.) In similar fashion it is easy to see that if the game begins with a vacancy anywhere on the board, the position can be transformed by Charosh's method to a single counter on the same cell. Again, this can always be done in actual play.

Is it possible to begin with a center vacancy and end with the last counter on 45? No, it is not. There is no way that Charosh's method can be used to transform the board to a lone counter on 45. To prove this we do not have to start with a full board. We can begin with the single counter on 44 (which we know to be a possible ending) and determine how this position can be transformed to other positions with a lone counter. Thus: The counter on 44 can be removed and counters placed on 54 and 64 (because 44, 54, 64 form a triplet). The counters on 54 and 64 can in turn be taken away and replaced by a counter on 74. So a lone counter on 44 is "equivalent" to a lone counter on 74. We can put it this way: A single counter is equivalent to a single counter on any cell that can be reached by jumping over two cells in a straight line in any orthogonal direction. It is easy to see that 44 is equivalent only to cells 14, 47, 74, 41. These are the only cells on which it is possible to end a game that begins with a vacancy in the center. Practice bears this out. Any final jump that puts a counter in the center can be made in the opposite direction to put a counter in an equivalent cell. All five cells, therefore, can be reached in actual play—but no others.

Application of Charosh's method will reduce any position either to a single counter, two counters diagonally adjacent or no counters. The last cannot, of course, be reached in actual play; instead the game must end on a position equivalent to no pieces, such as three adjacent counters in a row, or two in a row with two spaces between them. It is not hard to show that any position is equivalent (transformable by Charosh's method) to its "inverse"—that is, to the same position with vacancies replaced by counters and counters by vacancies. For example, if counters are removed from two diagonally adjacent cells, say 37 and 46, the position is equivalent to an empty board with counters on those same two cells. Because there is no way to transform those two counters to a single counter, we know that it is not possible to start with vacancies at 37 and 46 and reduce the board to a single counter.

For anyone wishing to devise a new solitaire problem, Charosh's system can save endless hours of time spent in seeking solutions for impossible problems. Of course, once a problem is shown to be not impossible, the task of finding a solution remains. Sometimes a solution exists, sometimes it does not. In seeking a solution, Leibniz's method of working backward has one enormous advantage: using numbered counters and taking them in order makes it unnecessary to keep a record of each attempt. If the attempt succeeds, the numbers make it easy to reconstruct the sequence of the play.

In 1960 Noble D. Carlson, an engineer in Willoughby, Ohio, raised an interesting question: What is the smallest *square* solitaire board on which it is possible to start with a full board, except for a vacancy at one corner, and reduce the position to a single counter? Charosh's technique quickly shows that this is impossible on all squares except those with sides that are multiples of three. The 3 × 3 square, however, proves to be unsolvable. This leaves the 6 × 6 as the most likely candidate. (*See Figure 62.*) The solution, if there is

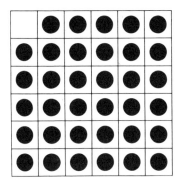

Figure 62
The 6 X 6 problem

one, will end on the corner cell left open at the start or on one of the three cells "equivalent" to it. (Let the vacancy be at cell 1, in the upper left-hand corner, and number the cells left to right. The three equivalent cells are 4, 19 and 22.)

Can it be done? Yes. Carlson himself found a twenty-nine–move solution ending on cell 22. What is wanted now is a solution beginning with a vacancy at 1 and ending at 1.

ADDENDUM

Readers called my attention to many early discussions of solitaire theory in which possibility tests, all more or less the same, are given. I have listed the more important references in the bibliography, including a recent report by J. D. Beasley of results obtained by a group of mathematicians (J. H. Conway, R. L. Hutchings and J. M. Boardman) at Cambridge University. Since Beasley's paper appeared, he and Conway have extended the theory even further, but their extensions have not yet been published. Sheldon B. Akers, Jr., a mathematician at the General Electronics Laboratory, Syracuse, New York, sent me his own procedure, equivalent to Charosh's method, by which a single number is assigned to any given solitaire position in such a way that "equivalent" positions have the same number.

Gary D. Gordon, a physicist at the RCA Astro-Electronics Products Division in Princeton, New Jersey, told me of a remarkable discovery that he had made some fifteen years earlier. The solution to every solitaire problem, on any board, that starts with only one vacant cell and ends with the last counter on that same cell, is reversible. That is, the jumps can be taken in reverse order to provide a new solution for the same problem. This should not be confused with Leibniz's method of working backward by starting with an empty board. The beginning position remains the same; only the order of jumps is reversed. Thus, in the sixteen-move solution provided by John Harris of Santa Barbara, California, for the 6 × 6 square, the reversed solution begins with 13-1, continues with 25-13, 27-25, and so on, taking in reverse order the jumps in the final eight-ball sweep. The result is a different solution with thirty-one moves. Davis points out that on full boards there are reverse solutions even if the initial and terminal cells are not the same. If a solution is found starting with cell *a* and ending on cell *b*, a reversal of the moves automatically provides a solution starting with *b* and ending on *a*.

Problems isomorphic with peg-solitaire problems are sometimes given as checker-jumping problems on a standard checkerboard. One of the oldest and best known of such puzzles begins with 24 checkers on the 24 black squares that are in the border, two cells wide, around the four sides of the board. Is it possible to reduce these checkers to one by jumping? Harry Langman discusses this problem in *Scripta Mathematica*, September 1954, pages 206–8, and in his book *Play Mathematics* (New York: Hafner, 1962), pages 203–6. An earlier discussion appeared in *Games Digest*, October 1938, and the problem goes back at least to 1900. It is easy to set up the isomorphic problem on a peg-solitaire field and test for possibility as explained by B. M. Stewart in his 1941 magazine article cited in the bibliography. It fails to pass. If, however, either of the two corner checkers is removed, there are many solutions.

Conway has informed me that the standard solitaire game, with center-hole vacancy, can be played to an unsolvable position in as few as four moves: jump into center, over center, into center, over center, the first and last moves being in the same direction. This is the shortest way to do it and is unique for four moves. In five moves one can reach two different unsolvable positions.

Bergholt, in his book on solitaire, asserted that it is possible to start with a corner vacancy on the standard board and conclude with a nine-ball sweep. He gave no solution. As far as I know, a solution to this difficult problem was first rediscovered by Harry O. Davis who gives an elegant eighteen-move solution in his 1967 article, cited in my list of references. Davis also shows in this article that no solution to the standard game, regardless of what cell is vacant at the start, can contain a chain longer than nine moves.

Davis, who has been mentioned many times in this chapter, first became interested in peg solitaire when he read about it in my column in 1962. Since then he has made enough fresh discoveries—extending possibility tests, developing techniques for obtaining minimum-move solutions and proving them minimal, creating and solving new problems, and even extending solitaire to three dimensions (which he calls "solidaire")—to make a sizable book. So far he has published only the one article listed. In recent years he has been collaborating with Wade E. Philpott, Lima, Ohio, who has done important work on the theory of peg solitaire both in its traditional orthogonal form and also on isometric (triangular) fields. (On isometric peg solitaire see my *Scientific American* columns for February and May, 1966.)

ANSWERS

For the first five problems, readers found shorter solutions than the ones I gave in *Scientific American*. I have here substituted minimal solutions, giving the names of those who sent such solutions.

Greek cross in six moves: 54-74, 34-54, 42-44-64, 46-44, 74-54-34, 24-44. (R. L. Potyok, H. O. Davis.)

Fireplace in eight moves: 45-25, 37-35, 34-36, 57-37-35, 25-45, 46-44-64, 56-54, 64-44. (W. Leo Johnson, H. O. Davis, R. L. Potyok.)

Pyramid in eight moves: 54-74, 45-65, 44-42, 34-32-52-54, 13-33, 73-75-55-53, 63-43-23-25-45, 46-44. (H. O. Davis.)

Lamp in ten moves: 36-34, 56-54, 51-53-33-35-55, 65-45, 41-43, 31-33-53-55-35, 47-45, 44-46, 25-45, 46-44. (Hugh W. Thompson, H. O. Davis.)

Inclined square in eight moves: 55-75, 35-55, 42-44, 63-43-
45-65, 33-35-37-57-55-53-51-31-33-13-15-35, 75-55, 74-54-56-
36-34, 24-44. (H. O. Davis.) Note the remarkable chain of
eleven jumps.

Wall: 46-44, 43-45, 41-43, 64-44-42, 24-44, 45-43-41.
This solves the problem. By continuing to play it is easy to
reduce the figure to four pieces on the corners of the central
3 by 3 square.

Square: 46-44, 25-45, 37-35, 34-36, 57-37-35, 45-25, 43-45,
64-44, 56-54, 44-64, 23-43, 31-33, 43-23, 63-43, 51-53, 43-63,
41-43. The finish is apparent: 15-35, 14-34, 13-33 on the left,
and the corresponding moves on the right, 75-55, 74-54, 73-
53. The puzzle is now solved. Four more jumps will leave
counters on the corners (36, 65, 52, 23) of an inclined square
—an unusually difficult pattern to achieve if one does not
know earlier positions.

Pinwheel: 42-44, 23-43, 44-42, 24-44, 36-34, 44-24, 46-44,
65-45, 44-46, 64-44, 52-54, 44-64. The position now has fourfold symmetry. It is completed: 31-33, 51-31, 15-35, 13-15,
57-55, 37-57, 73-53, 75-73. The final figure is a stalemate.

The shortest stalemate, starting with a full board and a vacant center cell, is reached in these six moves: 46-44, 43-45,
41-43, 24-44, 54-34, 74-54. The next-shortest stalemate is a
ten-move game.

Robin Merson, who works on satellite orbit determinations
at the Royal Aircraft Establishment in Farnborough, England, sent a simple proof that at least sixteen moves (a chain
of jumps counts as one move) are necessary in solving the
problem on the 6 × 6 square. The first move is 3-1, or its
symmetrical equivalent. This places a counter on each corner
cell. It is impossible for a corner piece to be jumped, therefore each corner piece must move (including the counter at
1, which must move out to allow a final jump into the corner). These four moves, added to the first, bring the total to
five. Consider now the side pieces on the borders between
corners. Two such pieces, side by side, cannot be jumped;
therefore for every such pair at least one counter must move.
On the left and right sides, and on the bottom, at least two
pieces must move to break up contiguous pairs. On the top
edge (assuming a 3-1 first move) one piece will suffice. This

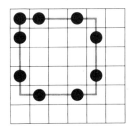 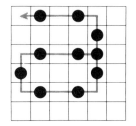

Figure 63

Eight-ball-sweep solution   Nine-ball-sweep solution

adds seven moves, carrying the total to twelve. Consider next the sixteen interior cells. A block of four (e.g., 8, 9, 14, 15) cannot be jumped until at least one man has moved. It is easy to see that a minimum of four interior pieces must be moved to break up all interior four-cell blocks. This brings the total of required moves to sixteen. Merson's shortest solution was eighteen. He wondered if the gap could be narrowed.

To my amazement, one reader, John Harris of Santa Barbara, California, came through with the ultimate—an elegant sixteen-move solution: 13-1, 9-7, 21-9, 33-21, 25-13-15-27, 31-33-21-19, 29-27, 16-28, 24-22, 18-16, 6-18, 36-24-12, 3-15-17, 35-33-21-23, 4-16-18-6-4, 1-3-5-17-29-27-25-13-1. Note that the final move is an "eight-ball sweep." The top illustration of Figure 63 shows the pattern just before this final move. In 1964 H. O. Davis found sixteen-move solutions with the initial vacancy at any cell of the 6 × 6 square.

The longest possible final chain is nine jumps. This was achieved by Donald Vanderpool of Towanda, Pennsylvania, at the close of an eighteen-move solution: 13-1, 9-7, 1-13, 21-9, 3-15, 19-21-9, 31-19, 13-25, 5-3-15, 16-4, 28-16, 30-28, 18-30, 6-18, 36-24-12-10, 33-21-9-11, 35-33-31-19, 17-15-13-25-27-29-17-5-3-1. The position before the final sweep is shown at the right of Figure 63.

Vanderpool also investigated rectangular boards with a vacant corner cell. He proved that every such board, including square boards, with one dimension of three cells or a multiple of three, has a solution except for the following boards:

1. One dimension of one cell (except for the 3 × 1).
2. One dimension of two cells.
3. The 3 × 3 square.
4. The 3 × 5 rectangle.

CHAPTER TWELVE

O

Flatlands

SATIRE often takes the form of fantasy in which human customs and institutions are caricatured by a race of nonhuman creatures or a society or world with its own peculiar standards or physical laws. Twice there have been notable attempts to base such satire on a society of two-dimensional creatures moving about on a plane. Neither attempt can be called a literary masterpiece, but from a mathematical point of view both are curious and entertaining.

Flatland (first published in 1884 and now, happily, available as a Dover paperback) is the earlier and better known of the two. It was written by Edwin Abbott Abbott, a London clergyman and school headmaster who wrote many scholarly books. The title page of the first edition bears the pseudonym of A. Square. The book's narrator is a square in the literal sense. He possesses a single eye at one of his four corners. (How he managed, without feet, to move over the surface of Flatland and how he managed, without arms, to write his book are left unexplained.)

Abbott's Flatland is a surface something like a map, over which the Flatlanders glide. They have luminous edges and an infinitesimal height along the vertical coordinate, or third dimension, but they are completely unaware of their height and have no power to visualize it. Society is rigidly stratified. At the lowest level are the women: simple straight lines with an eye at one end, like a needle. There is a visible glow from a woman's eye, but none from her other end, so that she can make herself invisible simply by turning her back. If a male Flatlander inadvertently collides with a lady's sharp posterior, the encounter can be fatal. To avoid such mishaps, women are required by law to keep themselves visible at all times by a perpetual wobbling of their rear end. Among ladies married to men of high rank this is a "rhythmical" and "well-modulated undulation." Lower-class females try to imitate it but seldom achieve anything better than "a mere monotonous swing, like the ticking of a pendulum."

Soldiers and workmen of Flatland are isosceles triangles with extremely short bases and sharp points. Equilateral triangles constitute the middle class. Professional men are squares and pentagons. The upper classes start as hexagons, and the number of their sides increases with their rank on the social ladder until their figures are indistinguishable from circles. The circles, who top the hierarchy, are the administrators and priests of Flatland.

In a dream the square narrator visits Lineland, a one-dimensional world, where he fails to convince the king of the reality of two-dimensional space. In turn the square receives a visitor from Spaceland—a sphere who initiates him into the mysteries of three-space by lifting him above Flatland so that he can look down into the interior of his pentagonal house. When he returns to Flatland, the square tries to preach the gospel of three-space, but he is thought mad; he is arrested for his views and is in prison as the tale ends.

The sphere had entered Flatland by moving slowly through the plane until his cross section reached a plane figure of maximum area. It is easy to see that this section is a circle with a radius equal to the radius of the sphere. Suppose that instead of a sphere a cube had entered Flatland. What is the maximum area of a plane cross section that a cube of unit side could attain? The cube can, of course, tip his body at any angle as he crosses the plane.

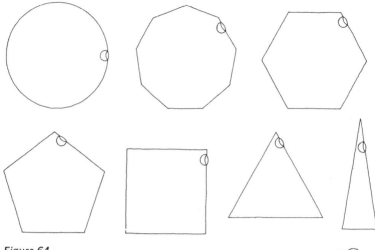

Figure 64
One-eyed Flatlanders, in order of social rank

A much more ambitious work of two-dimensional fiction than Abbott's—indeed, a full-blown 181-page novel—was Charles Howard Hinton's *An Episode of Flatland*, published in London in 1907. Hinton was the son of James Hinton, a prominent London ear surgeon who was a friend of George Eliot's and the author of *The Mystery of Pain* and other widely read books. Young Charles studied mathematics at Oxford, married Mary Boole (one of the five daughters of George Boole, the logician) and settled in the United States. He taught mathematics at Princeton University and at the University of Minnesota. When he died, in 1907, he was an examiner in the United States Patent Office.

A long obituary in the New York *Sun* (May 5, 1907, page 8) was written by Gelett Burgess of purple-cow fame. Burgess recalls an occasion when his friend Hinton was attending a football game and a stranger tried to snatch a chrysanthemum from his lapel. Hinton picked up the man, tossed him over a nearby fence. In 1897 Hinton was in the news with his invention of an automatic baseball pitcher. (For details, see *Harper's Weekly*, Vol. 41, March 20, 1897, pages 301–2.) It shot balls with charges of gunpowder and could be adjusted to produce a pitch of any desired speed or curve. The Princeton team practiced with it for a while, but after a few accidents the batters were afraid to face it.

Hinton was best known as the author of books and articles on the fourth dimension. He developed a method of building models of four-space structures (in three-space cross sections), using hundreds of small cubes, labeled and colored in a manner detailed in his two most important books, *The Fourth Dimension* and *A New Era of Thought*. By working with these cubes for many years, Hinton maintained, he actually learned to think in four dimensions. He taught the method to his sister-in-law Alicia Boole when she was eighteen. Although the girl had no formal schooling in mathematics, she soon developed a remarkable grasp of four-space geometry and later made significant discoveries in the field. (See H. S. M. Coxeter, *Regular Polytopes* [New York: Macmillan, 1948], pages 258–59, for the story of her unusual career.) The wife of Hinton's son Sebastian is Carmelita Chase Hinton, founder and retired head of the Putney School in Vermont.

In constructing his Flatland, which he called Astria, Hinton took a more ingenious approach than Abbott. Instead of allowing his creatures to wander at will over the surface of a plane, he stood them upright, so to speak, on the rim of an enormous circle. If you place coins of various sizes on a table and slide them about, you will find it easy to imagine a flat sun around which flat circular planets orbit. Gravity behaves as it does in our space, except that on the plane its force naturally varies inversely with the distance instead of with the square of the distance.

The planet Astria is depicted in Figure 65. The direction (indicated by the arrow) in which it rotates is called east,

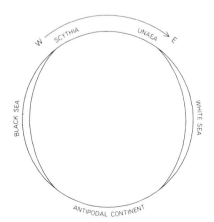

Figure 65
Charles Hinton's
two-dimensional planet, Astria

Figure 66
Home life among the Astrians

the opposite direction west. There are no north and south, only up and down. The Astrians' bodies have a complex structure, but to avoid going into anatomical details Hinton represents them schematically as right triangles in the manner shown in Figure 66. Like Abbott's Flatlanders, the Astrians have only one eye. (Apparently neither writer considered the possibility of introducing two-dimensional vision involving a pair of eyes each with a one-dimensional retina.) Unlike the Flatlanders, they have arms and legs. To pass each other, two Astrians must of course go under or over each other, as would two acrobats on a tightrope. All male Astrians are born facing east, all females facing west. They keep this orientation until they die because there obviously is no way for an Astrian to "turn over" to become his mirror image. To see behind him an Astrian must bend backward, stand on his head or use a mirror. The mirror method is the most convenient; for this reason Astrian houses and buildings are well supplied with mirrors. To kiss his son a father must hold the boy upside down.

The inhabited region of Astria was originally divided between the civilized Unaeans in the east and the barbarian Scythians in the west. The Scythians had one great advantage in warfare: their male warriors could strike the Unaeans from behind, whereas the Unaeans could retaliate only by the awkward method of hitting backward. As a result the Scythians drove the Unaeans eastward until they were squeezed into a narrow territory bordering the White Sea.

The Unaeans were saved from extinction by the rise of science. Their astronomers, observing eclipses and other phenomena, became convinced of the roundness of their planet. A study of tides in the White Sea enabled them to deduce the existence of an antipodal continent. A select band of Unaeans sailed over the White Sea and crossed the new continent in a 100-year march during which each tree along the route had to be climbed over or cut down. Sons and daughters who survived the ordeal then built new ships to cross the Black Sea. The Scythians, taken by surprise, were quickly overwhelmed because now it was the Unaean men who could attack from the rear! World government was established; an era of peace had begun. All this is background history to set the stage for the novel.

I will spare the reader the details of the book's melodramatic two-dimensional plot. It is in the tradition of early socialist fantasies, attacking plutocracy in the name of an altruistically planned society. There is a rather flat love affair involving Laura Cartright, beautiful daughter of the rich, powerful Secretary of State, and Harold Wall, her handsome (in a plane sort of way) proletarian suitor. Central to the plot is an ominous note of doom: The close approach of Ardaea, another planet, is expected to change Astria's orbit to an ellipse so eccentric that the climate will become alternately too hot and too cold to support life. The government begins a vast shelter program, excavating deep subterranean chambers and stocking them with provisions for the survival of the upper class.

The dreaded fate is averted by the mathematical theories of Laura's uncle, Hugh Miller, an eccentric old bachelor who lives on Lone Mountain. Miller (a thinly disguised Hinton) is the only man on the planet who believes in a third dimension. He has convinced himself that all objects have a slight thickness along a third coordinate; that they slide about over the smooth surface of what he calls the "alongside being." By working with models he has been able to awaken in himself a sense of three-space forms. He has come to understand that he is actually a three-space man directing a corporeal two-space body.

"Existence itself stretches illimitable, profound, on both

sides of that alongside being," Miller says in an eloquent address to the leaders of Astria. "Realize this . . . and never again will you gaze into the blue arch of the sky without an added sense of mystery. However far in those never-ending depths you cast your vision, it does but glide alongside an existence stretching profound in a direction you know not of.

"And knowing this, something of the old sense of the wonder of the heavens comes to us, for no longer do constellations fill all space with an endless repetition of sameness, but there is the possibility of a sudden and wonderful apprehension of beings, such as those of old time dreamed of, could we but . . . know that which lies each side of all the visible."

If there were some mechanical means of touching or latching on to the surface of the "alongside being," it would be possible to alter Astria's course in such a way that it might escape the influence of the approaching planet. There is no such method. But since the true self is three-dimensional, it may possess such power. The old man proposes a mass effort at what is today called psychokinesis, or PK—the power of thought to influence the motion of objects. The plan is carried out successfully. A concerted PK effort on the part of everybody alters Astria's orbit just enough to avert catastrophe. Science, armed with the new knowledge of three-space, begins a great leap forward.

It is amusing to speculate on two-dimensional physics and the kinds of simple mechanical devices that would be feasible in a flat world. Hinton points out elsewhere (in an essay on "A Plane World") that houses on Astria cannot simultaneously have more than one opening. When the front door is open, the windows and back door must be closed to keep the house from collapsing.

A tube or pipe of any kind is impossible: how could its sides be joined without obstructing the passageway? Ropes cannot be knotted. (It has been rigorously proved that closed lines knot only in three-space, the surface of a sphere knots only in four-space, the surface of a hypersphere only in five-space, and so on.) Hooks, levers, couplings, tongs and pendulums can be used, as can wedges and inclined planes. Wheels with axles are out of the question. A crude gear transmission might be made possible by partially encasing each wheel in a curved rim. Methods can be worked out for rowing ships; airplanes would have to fly like birds by flapping wings. Flatfish should have little difficulty paddling through the water with properly shaped fins. Liquor could be kept in bottles and poured into glasses but no doubt would taste flat. Heavy objects can be transported by rolling them along on circles much as a three-space object can be rolled over cylinders.

This Astrian method of moving objects introduces a delightfully bewildering problem sent to me recently by Allan B. Calhamer, a reader in Billerica, Massachusetts. Figure 67 shows a loaded Astrian flatcar, 30 feet long, that is being moved along a straight track by means of three circles. The

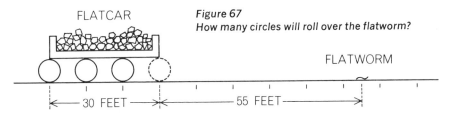

FLATCAR

Figure 67
How many circles will roll over the flatworm?

FLATWORM

←— 30 FEET —→ ←———— 55 FEET————→

circles are at all times exactly 10 feet apart from center to center. As soon as the position shown is reached, the rear circle is picked up by an Astrian at the rear and tossed to a companion in front, who places it at the spot shown by the broken line. The flatcar is pushed forward over the three circles, which roll along the track, until the wheels are once again in the position shown. The back circle is tossed to the front as before and the procedure is repeated as often as necessary.

The flatcar is being moved off the page to the right. Exactly 55 feet in front of the point at which the dotted circle touches the track is a flatworm. Assuming that the worm does not move, how many circles will roll over it?

The reader is urged to try to solve the problem first in his head. Next, check your answer with pencil and paper; finally, compare it with the answer at the close of this chapter. For those who would like to do a bit more homework, generalize for n equally spaced wheels. Surprisingly, it is not necessary to know the size of the wheels.

ADDENDUM

In describing Flatland I said that no tunnels were possible, but this is not strictly true. Gregory Robert, North St. Paul, Minnesota, wrote to say that the roof of a Flatland tunnel could be supported by a series of doors, each hinged at the top. A Flatlander could walk through such a tunnel, opening one door at a time while the roof remained supported by the other doors. There would have to be a mechanism to prevent all doors from being opened at once.

"The Fourth Dimension: An Efficiency Picture," Chapter 12 of Fletcher Durell's *Mathematical Adventures* (Boston: Bruce Humphries, 1938) contains some amusing speculations about inhabitants of Thinland, a region similar to Hinton's Flatland. Binocular vision is achieved by two eyes, one on the forehead, one on the chin. A long neck permits a Thinlander to tilt his head backward and upside down to see behind him. When male and female Thinlanders have to pass each other, the rule is that the man lies down to let the woman walk over him.

In addition to these mechanical difficulties of life on the plane, mention should also be made of the problem of designing a brain in view of the topological limitations of planar networks. An animal brain as we know it demands a fantastically complex three-space network of nerve filaments impossible to achieve on the plane without self-intersection. The difficulty is not, however, as formidable as it seems, for one can imagine self-intersecting networks along which electrical impulses travel across intersections without, so to speak, turning corners.

For information on Boole's wife and five daughters, and their remarkable descendants, the reader is referred to Norman Gridgeman's article "In Praise of Boole," in *The New Scientist*, No. 420, December 3, 1964, pages 655–57. Boole's wife, Mary, in the sixty years after her husband's death, "continually wrote about and preached boolery in a dozen fields," writes Gridgeman, "including theology and ethics. She became almost obsessed with the mystique of algebraic symbolism and the roles of zero and unity. As late as 1909 she put out a book entitled *The Philosophy and Fun of Algebra* in which she urged 'those who wish . . . to get into right relations with the Unknown' to create their own algebras on boolean principles."

Howard Everest Hinton, a grandson of Charles Hinton and Mary, the oldest daughter of Boole, is a well-known British entomologist. The story of two of Charles's other grandchildren, William Hinton and his sister Joan, a physicist, is told in *Time,* August 9, 1954, page 21. Both became enthusiastic supporters of Red China. The son of Boole's second daughter, Margaret, is Geoffrey Taylor, a Cambridge mathematician. The story of Alicia, the third daughter, has already been briefly told. The fourth daughter, Lucy, became a professor of chemistry at the Royal Free Hospital, London. Ethel Lillian, the youngest daughter, married Wilfrid Voynich, a refugee Polish scientist. In her youth she wrote *The Gadfly,* a bitterly anti-Catholic novel about political revolution in Italy. It became one of the all-time best sellers in Russia and, more recently, in China. After the First World War the Voyniches moved from London to Manhattan, where Ethel died in 1960 at the age of ninety-six. "Modern Russians are constantly amazed," writes Gridgeman, "that so few Westerners have heard of E. L. Voynich, the great English novelist."

ANSWERS

The problem of slicing a cube to obtain a plane section of maximum area is answered as shown in Figure 68. The shaded section is a rectangle with an area of $\sqrt{2}$, or 1.41+. (The problem was posed by C. Stanley Ogilvy and answered by Alan R. Hyde in *American Mathematical Monthly,* Vol.

63 [1956], p. 578.) It is possible to slice a cube so that the section is a regular hexagon, but the area is only 1.29 +.

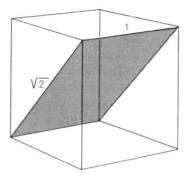

Figure 68
Answer to cube problem

The answer to the flatcar problem is that only one circle will roll over the flatworm. When there are n equally spaced circles and n is even, the number of circles that roll over a flatworm placed anywhere on the track (except at spots where a circle goes directly on the worm) is $n/2$. When n is odd, the situation is more complex. The track ahead of the front circle must be divided into segments that are each equal in length to the spacing between two adjacent circles. A worm on any alternate segment, beginning with the segment immediately ahead of the front circle, will be run over by $n/2 + 1/2$ circles. A worm on any of the other alternate segments will be run over by $n/2 - 1/2$ circles. Again, one assumes that the worm is not at a spot where a circle is placed directly on top of it; or, as the mathematician would say, one ignores "boundary conditions."

Readers who solved the problem will have observed that the flatcar moves twice as fast relative to the ground as a wheel rolling beneath it, so that for every distance x traveled by a wheel, the flatcar goes a distance of $2x$. The same mechanical principle is involved in the operation of elevator doors; one of the half-doors slides twice as fast and twice as far as the other.

CHAPTER THIRTEEN

O

Chicago
Magic Convention

EVERY SUMMER, usually in July, several thousand members of the imaginary Brotherhood of American Magicians descend on a Middle Western hotel for their annual convention. This year it was the Sherman Hotel, at the northwest corner of Chicago's Loop. For three days and nights the hotel lobby was a phantasmagoria of riffling cards, clicking coins, cut and restored ropes, fluttering doves, vanishing bird cages and even one or two levitated ladies.

I attended the conclave partly because magic is my principal hobby, partly in search of offbeat material for my *Scientific American* column. Many professional mathematicians are amateur conjurers and many conjurers have a lively interest in mathematics. The result is mathemagic, surely the most colorful of all the branches of recreational mathematics.

On the mezzanine floor about twenty magic dealers had set up booths for the purpose of hawking their wares. I paused in front of the booth where the Great Jasper (a Chicago magic dealer who performs under that name) was demonstrating a large-size version of what magicians call "tumble rings." Thirty steel rings are linked together in the curious manner shown in Figure 69. To operate the tumble rings, first hold the top ring of the chain in the left hand. Directly below the top ring is a pair of rings. With the right thumb and index finger take the *back* of the ring on the right exactly as shown in the illustration. When the ring held in the left hand is released, it appears to tumble from ring to ring all the way down the chain, finally linking itself to the bottom ring.

To repeat the effect, hold what is now the top ring in the right hand. With the left thumb and index finger hold the *front* of the ring on the left of the pair that links through the top ring. When the top ring held in the right hand is released, it tumbles down the chain as before.

"Do you suppose any of my readers could make a set of these rings?" I asked.

"Why not?" said Jasper. "Five-and-dime stores sell steel key rings. With thirty key rings and a strong thumbnail you can make a set of tumble rings in about twenty minutes. But don't tell any of the other dealers I said so."

Jasper was right. Key rings of the familiar coiled type make excellent tumble rings. To save your thumbnail, use a nail file to pry open the ends of the coils.

Figure 69
The tumble rings

A twist of the blade will keep a ring open until another ring can be slipped into the gap. The least confusing procedure is to start with the top ring, hanging it on a projection, then work down ring by ring, following the illustration. The rings tumble smoothly, with a pleasant clicking rhythm, unless you have made a mistake in the linkage.

While Jasper and I were chatting, Fitch Cheney, a mathematician at the University of Hartford, came over and joined us. "If you're interested in linkage effects," he said to me, "I've invented a new one that your readers might like."

From his pocket Cheney pulled a long piece of soft rope. Jasper and I each took an end, then with the index finger of our free hands we bent the rope into the shape shown at *a* in Figure 70. Cheney tied a silk handkerchief tightly around the rope by making a single knot as shown at *b*. Both ends

Figure 70
Steps in performing Fitch Cheney's rope-and-handkerchief trick

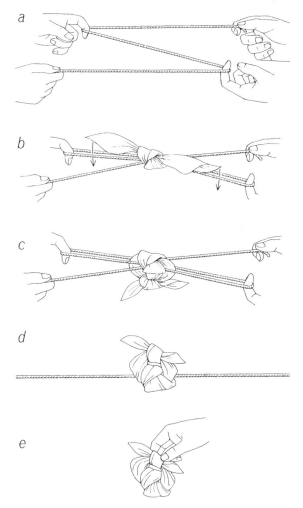

of the handkerchief were then tucked down through a loop, as indicated by the arrows, and the ends were tied twice below the rope to make a secure square knot [c].

"Please release the loops you are holding with your index fingers," Cheney said, "and remove all the slack from the rope by pulling it straight." We did so, with the result shown at d. Cheney rotated the knotted silk 180 degrees to bring the square knot to the top.

"It's a strange thing," he said. "Although that handkerchief has been knotted tightly around the rope, the rope is now *outside* the closed curve formed by the cloth." He took hold of the knotted handkerchief, lifted it up and off the rope as shown at e! The effect is self-working if you follow the illustrations carefully.

The hotel's cocktail lounge before the dinner hour was noisy with prestidigitators. At the bar I ran into my old friend "Bet a Nickel" Nick, a blackjack dealer from Las Vegas who likes to keep up with the latest in card magic. The nickname derives from his habit of perpetually making five-cent bets on peculiar propositions. Everybody knows his bets have "catches" to them, but who cares about a nickel? It was worth five cents just to find out what he was up to.

"Any new bar bets, Nick?" I asked. "Particularly bets with probability angles?"

Nick slapped a dime on the counter beside his glass of beer. "If I hold this dime several inches above the top of the bar and drop it, chances are one-half it falls heads, one-half it falls tails, right?"

"Right," I said.

"Betcha a nickel," said Nick, "it lands on its *edge* and stays there."

"O.K.," I said.

Nick dunked the dime in his beer, placed it against the side of his glass and let it go. It slid down the straight side, landed on its edge and stayed on its edge, held to the glass by the beer's adhesion. I handed Nick a nickel. Everybody laughed.

Nick tore a paper match out of a folder, marked one side of the match with a pencil. "If I drop this match, chances are fifty-fifty it falls marked side up, right?" I nodded. "Betcha a nickel," he went on, "that it falls on its edge, like the dime."

"It's a bet," I said.

Nick dropped the match. But before doing so, he bent it into the shape of a V. Of course it fell on its edge and I lost another nickel.

Someone in the crowd took a small plastic top from his pocket. "Have you seen these 'tippy tops' that the dealers are selling? I'll bet you a nickel that if you spin it, it will turn upside down and spin on the tip of its pin."

"No bet," said Nick. "I bought a tippy top myself. But I'll tell you what I'll do. You spin the top clockwise. I'll bet *you* a nickel you can't tell me now in what direction it will be spinning after it flips over."

The man with the top pursed his lips and mumbled: "Let's see. It goes clockwise. When it turns over, it will have to keep spinning the same way. Obviously it can't stop spinning and start again in the other direction. But if the ends of its axis are reversed, the spin will be reversed when you look down on the top. In other words, after the top flips over it should be spinning counterclockwise."

He gave the top a vigorous clockwise spin. In a moment it turned upside down. To everybody's vast astonishment *it was still spinning clockwise when one looked down on it. (See Figure 71.)* If the reader will buy a tippy top (they are sold in many dime stores and toy shops), he will discover that this is indeed what happens. As a particle physicist might say, the top actually alters its parity as it turns over. It becomes its own antitop or mirror image!

After the banquet and evening show, conventioners clustered in various hotel rooms to gossip, swap secrets and talk magic. I finally located the room in which the mathemagicians were in session. A friend from Winnipeg, Mel Stover,

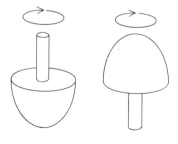

Figure 71
A clockwise spun "tippy top"
[left] is still spinning clockwise
after it turns upside down
[right]

was explaining how the binary system could be applied to a familiar method of revealing a chosen card.

In many card tricks the selected card is disclosed when the spectator is handed a small packet of cards and asked to shift the top card to the bottom of the packet, deal the next card to the table, shift the next card to the bottom, deal the next to the table, and so on, until only one card remains. It proves to be the selected card. At what position in the packet must this card originally be placed so that it will become the last card? The position will vary, of course, with the number of cards in the packet. It can be determined by experiment, but for large packets experimenting is tedious. Fortunately, Stover explained, the binary system provides a simple answer.

This is how it is done. Express the number of cards in the binary system, shift the first digit to the end of the number, and the resulting binary number will indicate the position that the chosen card should be in from the top of the original packet. For example, suppose an entire deck of 52 cards is used. The binary expression for 52 is 110100. We move the first digit to the end: 101001. This new number is 41, therefore the chosen card must be the 41st card from the top of the deck.

What size packets can be used if we want the top card of the packet to be the card that remains? The binary number for the position of the top card is 1, so we must use packets with binary numbers of 10, 100, 1000, 10000 . . . (in decimal notation packets of 2, 4, 8, 16 . . . cards). If we want the *bottom* card of the packet to be the remaining card, then the binary numbers of the packets must be 11, 111, 1111, 11111 . . . (or 3, 7, 15, 31 . . . cards).

Is it possible for the second card from the top of a packet to be the remaining card? No. In fact, no card at an *even* position from the top can ever be the remaining card. The position of the chosen card, expressed as a binary number, must end in 1 (because after the first digit, which must be 1, is moved to the end it forms a number ending in 1). All binary numbers ending in 1 are *odd* numbers.

Victor Eigen, whose tricks were discussed in *New Mathematical Diversions from Scientific American* (New York: Simon and Schuster, 1966), took the floor to demonstrate a remarkable new card trick that involves the coding of information. "I want to explain in advance exactly what I intend to do," he said. "Anyone may shuffle his own deck of cards and from it select any five cards. From those five he must choose one. I am allowed to arrange the remaining four cards in any order I please. These four cards, squared into a packet and all face down, are to be taken to my hotel room by whoever selected the card. My wife is in the room, waiting to assist in the trick. The person carrying the packet will knock three times on the door, then push the packet of four cards, held face down, under the door. No words will be spoken by either person. My wife will examine the packet and name the selected card."

I asked permission to do the selecting. The procedure was carried out exactly as Eigen had directed. I took five cards from my own deck and selected from them the six of spades. Eigen did not touch the cards. He wanted to rule out the possibility that he might mark them in some way and so provide additional information. Moreover, most cards have backs that vary in minute details when turned upside down. By taking advantage of these "one-way backs" (as magicians call them) it would be possible to arrange the cards in a pattern—some turned one way, some the other—that would convey a large amount of information. If the cards had been placed in a container of some sort, say an envelope, still more information could be coded. For example, the cards could be put in the envelope either face up or face down, the envelope could be sealed or left unsealed, and so on. Even the choice of a container or no container could convey information. Had Eigen been given the privilege of picking someone to take the cards to his wife, this choice also could be used as part of the code. He could select a person with dark or light hair, married or unmarried, last initial from A to M or M to Z, and so on. Of course his wife would have to observe in some way who delivered the cards. It was to rule out all these possibilities that Eigen had described the procedure in advance and had been careful not to touch the cards in any way.

After I had arranged the four cards in an order specified by Eigen, I asked for his room number and was about to leave when Mel Stover spoke up. "Hold on a minute," he said.

"How do we know that Eigen isn't sending information by the *time* he picks to send you to his room? By conversation he delays your leaving until the time is within a certan interval that is part of the code."

Eigen shook his head. "No time intervals are involved. If you like, wait awhile and let Gardner go whenever he wishes."

We delayed about fifteen minutes, watching with awe while Ed Marlo, a Chicago card expert, showed how a flawless series of eight faro shuffles would bring a full pack back to its original order. A faro shuffle—in England it is called a weave shuffle—is a perfect riffle shuffle in which single cards alternate from left and right halves, each half containing twenty-six cards. If the first card to fall is from the former bottom half, it is called an out-shuffle. If the first card is from the former top half, it is an in-shuffle. Eight out-shuffles or fifty-two in-shuffles will restore the deck's original order. Only the most skillful card hustlers and magicians can execute such shuffles rapidly and without error. In recent years many articles analyzing the faro shuffle in the binary system have been published in both magic and mathematical journals. Ed Marlo has published two books about the shuffle and the brilliant mathematical card tricks that can be based on it.

After Marlo's demonstration I carried my packet of four cards to Eigen's room, knocked three times, pushed the facedown packet under the door. I heard footsteps. The packet was pulled out of sight. A moment later Mrs. Eigen's voice said: "Your card is the six of spades."

Exactly how did Eigen convey this information to her?

ADDENDUM

I have been unable to learn the origin of the tumble rings or even the approximate time they were invented. They are mentioned in R. M. Abraham's *Winter Nights Entertainments,* a British book published in 1932, but are undoubtedly much older. The chain is sometimes made with rings of two colors so that if you drop, say, a red ring, you see what appears to be a red ring tumble down and hang at the bottom. If the magician starts with a separate red ring palmed in one hand, a green ring in the other, he can apparently cause a ring of each color to tumble, and each time appear to catch and remove the ring as it falls off the lower end.

E. A. Brecht, Chapel Hill, North Carolina, found a good procedure for forming the chain. He starts with a 1-2-1-1 chain of simply linked rings. The bottom ring is then linked in the proper manner with the single ring above it to make a 1-2-2 chain. A 2-ring chain is added to make a 1-2-2-1-1 chain, then the bottom ring is again linked to the one above it in the proper manner. Another 2-ring chain is added, and this procedure repeated as often as desired.

For explanations of the tippy top, see C. M. Braams, "The Symmetrical Spherical Top," *Nature,* Vol. 170, No. 4314 (July 5, 1952) ; C. M. Braams, "The Tippe Top," *American Journal of Physics,* Vol. 22 (1954), page 568; and John B. Hart, "Angular Momentum and Tippe Top," *American Journal of Physics,* Vol. 27, No. 3 (March 1959), page 189.

The binary method that I gave for determining the position of a card in a packet of n cards (so that it will be the last card when one follows the procedure of alternately dealing a card to the table and placing a card under the packet) was published by Nathan Mendelsohn in *American Mathematical Monthly,* August-September, 1950. An equivalent way of calculating the position had long before been known to magicians: simply take from n the highest power of 2 that is less than n, and double the result. This gives the card's position if the first card is dealt to the table. If the first card is placed beneath the packet, 1 is added to the result. (If n is itself a power of 2, the card's position is on top of the packet if the first card goes beneath, on the bottom of the packet if the first card is dealt.)

The earliest published trick I know of that exploits this formula is Bob Hummer's "The Great Discovery," a printed sheet of instructions published by Kanter's Magic Shop, Philadelphia, in 1939. Since then, dozens of ingenious card tricks using the principle have been published, with new ones still appearing in the literature. John Scarne, the card and gambling expert, in 1950 published a pamphlet called *Scarne's Quartette,* explaining four tricks using the principle. (They later appeared in Bruce Elliott's *The Best in Magic* [New York: Harper, 1956], pages 116–20.) Here, from one of Scarne's four tricks, is a simple handling that shows how the principle can be cleverly concealed.

Someone shuffles a deck and hands it to you. Fan the deck, faces toward you, and state that you will determine in advance a card that will be selected. Note the top card of the deck and write its name on a slip of paper that you put aside without letting anyone see what you have written. Assume that the card is the two of hearts.

The deck is held face down in your left hand. Ask a spectator to give you any number from 1 to 52, but preferably above 10 to make the trick more interesting. Suppose he says 23. Mentally subtract the highest power of 2 you can, in this case, 16, to get 7. Twice 7 is 14. Your task now is to get the top card, the two of hearts, to the fourteenth position in a packet of 23 cards. This is done as follows. Count the cards singly by taking them from the top of the deck with your right thumb. This reverses the order of the cards. After counting 14, pause and say (as though you had forgotten), "What number did you give me?" When he tells you it was 23, nod, say "Oh, yes—twenty-three," and continue counting. Now, however, you take the cards from the deck by pushing them to the right with your left thumb and sliding each card *under* the packet in your right hand. Thus when you have counted 23 cards, the two of hearts has subtly been placed in the fourteenth position. Your pause and question breaks the counting into two parts, and no one is likely to notice that the two counting procedures are not the same. Hand the packet of 23 cards to the spectator with the request that he deal the first card to the table, the next one to the bottom of the pile in hand, the next to the table, and so on until a single card remains. It will, of course, be the card you predicted.

Sam Schwartz, a Manhattan attorney, published this presentation in 1962. I give it in slightly simplified form. Take from the deck a packet of 4, 8, 16, or 32 cards; for example, let us use 16. Turn your back and ask a spectator to remove a small packet of cards (it must be less than 16) from the deck and hold them in his hand without letting you know how many he took. Let n be the number he holds. Fan your packet of 16 cards, faces toward spectator, and ask him to remember the nth card from the top—without, of course, letting you know the value of n or the name of the card. Square up the cards you are holding and have him place his packet of cards on top. This automatically places the chosen card in the $2n$th position in a packet of $16 + n$ cards, therefore when the cards are dealt one to the table, then one to the bottom of the pile in hand, and so on, the selected card will be the one that remains. Instead of handing the packet immediately to the spectator, however, Schwartz puts it behind his back and says he will adjust the cards to put the chosen card in the desired position. Actually, he does nothing whatever behind his back. This is done, he writes, only "to conceal the fact that the trick is self-working."

Ronald Wohl, a chemist at Rutgers University who under the pseudonym of "Ravelli" has published many original mathematical tricks of great subtlety, has given me permission to describe the following unpublished trick, which has an effect similar to the preceding one and which he worked out independently at about the same time that Schwartz worked out his trick. After a packet of 2^n cards, say 32, has been shuffled by a spectator, he is asked to think of any number from 1 to 15 and put that number of cards into his pocket while the magician's back is turned. The performer takes the remaining cards and deals them face down into a pile, showing the face of each card as it is dealt. The spectator notes in his mind the card corresponding to his thought-of number. After all the cards are dealt (the dealing of course reverses their order) the packet is handed to a second spectator, who is given instructions for doing the "under-down" deal (first card under, next to the table, and so on) until one card—the chosen one—remains.

A different handling of essentially this same trick has been suggested by George Heubeck, a New York City card expert. A packet of 2^n cards is shuffled by a spectator, who deals it into two side-by-side piles, stopping whenever he wishes, provided that each pile has the same number of cards.

He is given a choice of either of the two piles or the packet in his hands. If he chooses a table pile, he notes its top card, then drops on that pile the cards in his hands. The enlarged pile is picked up and the chosen card disclosed by an under-down deal. If he chooses the packet in his hands he looks at the *bottom* card of that packet, drops it on either pile, picks up the cards, and finds the chosen card by a *down-under* deal.

The problem of determining the card's position in such tricks is a special case of a more general problem known to recreational mathematicians as the Josephus problem. It has been the basis of many old puzzles. A group of men stand in a circle. All but one are to be executed. The executioner starts counting round and round the circle, executing every nth man, until only one man remains. The last man is given his freedom. Where should a man stand in order to escape execution? When $n = 2$, we have the card situation. For a history of the Josephus problem and some of its ramifications, see W. W. Rouse Ball, *Mathematical Recreations and Essays*, revised edition (New York: Macmillan, 1960), pages 32–36.

The earliest reference I have on the five-card problem is Wallace Lee, *Math Miracles* (1950), Chapter 14, which explains a trick of Fitch Cheney's similar to the one I described. The difference is that in Cheney's version the magician is allowed to decide which of the five cards is to be the selected one. The problem of coding the fifth card when it is chosen by the spectator was first given, I believe, by "Rusduck" in the third issue (June 1957) of his obscure little magazine, *The Cardiste*. Issues 4 and 5 (September 1957 and February 1958) give two imperfect methods of doing the trick. (There are, of course, no "perfect" methods.) Further suggestions are supplied by Tom Ransom in the Canadian magic magazine *Ibidem*, No. 24 (December 1961), page 31, and many other methods have since appeared in other magic magazines.

ANSWERS

Since none of the four cards can be the selected card, it is necessary only to code the name of one of 48 cards. The magician and assistant have agreed on an order for all 52 cards, so that each card can be assigned a number, from 1 to 52, in the agreed-on hierarchy. The four cards that carry the code will then represent four numbers that can be designated A, B, C, D in order of rank. These four cards can be arranged in 24 different ways, exactly one half of 48. The 48 cards (one of which must be coded) are thought of as ordered according to the ranks of their assigned numbers, then divided in half, half consisting of the 24 lower cards, the other half consisting of the 24 higher cards. Suppose the chosen card is the seventeenth card in the "low" group. The number 17 can be communicated by the ordering of the four cards, but one additional signal is needed to indicate whether it is the seventeenth card in the "low" or the "high" group.

The problem that remains, then, is how to communicate this final yes-no signal. It cannot be communicated by the ordering of the four cards. The problem was stated in such a way as to rule out various other methods that suggest themselves, such as marks on the cards, the choice of the person who takes the cards to the assistant, the use of a container for the cards, the procedure to be followed, the time at which the cards are taken to the assistant and so forth.

One subtle loophole was *not* ruled out: the hotel room in which Mrs. Eigen waited. The Eigens had taken *two* rooms, adjoining and connecting. Victor Eigen did not give the number of his hotel room until after the card had been selected. He arranged the four cards to code a position from 1 to 24, then transmitted the final clue—whether the high or the low group was involved—by choosing one of his two rooms. Mrs. Eigen simply went to the door at which she heard knocking. This information, combined with the four-card code, was sufficient to pinpoint the selected card.

One reader in Manhattan, Robert S. Erskine, Jr., summed the situation up neatly with the following quatrain:

> *Two doors, two wives, or other plan, sir,*
> *Our friend must have, though necromancer,*
> *The cards alone, to girl or man sur-*
>   *render only half the answer.*

CHAPTER FOURTEEN

O

Tests of Divisibility

A DOLLAR BILL that I have just taken from my wallet bears the serial number 61671142. A schoolboy could say at once that this number is exactly divisible by 2 but not by 5. Is it divisible—from now on the word will be used to mean divisible without a remainder—by 3? By 4? By 11? Few people, including many mathematicians, know all the simple rules by which large numbers can be tested quickly for divisibility by numbers 1 through 12. The rules were widely known during the Renaissance, before the invention of decimals, because of their usefulness in reducing large-number fractions to lowest terms. Even today they are handy rules for anyone to know. For a devotee of digital puzzles the following rules are indispensable.

To test for 2: A number is divisible by 2 if and only if the last digit is even.

To test for 3: Sum the digits. If the result is more than one digit, sum again and continue until one digit remains. This final digit is called the digital root of the number. If it is a multiple of 3, the number is divisible by 3. If it is not a multiple of 3, its excess over 0, 3 or 6 is the same as the remainder when the original number is similarly divided. Example: The serial number of the bill has a digital root of 1. Therefore when the number is divided by 3, the remainder will be 1.

To test for 4: A number is evenly divisible by 4 if and only if the number formed by its last two digits is divisible by 4. (This is easy to understand when you reflect on the fact that 100 and all its multiples are evenly divisible by 4.) The dollar bill's serial number ends in 42. Because 42 has a remainder of 2 when divided by 4, the serial number, when divided by 4, will have a remainder of 2.

To test for 5: A number is divisible by 5 if and only if it ends in 0 or 5. Otherwise the last digit's excess over 0 or 5 equals the remainder.

To test for 6: Test for divisibility by 2 and 3, the factors of 6. A number is divisible by 6 if and only if it is an even number with a digital root divisible by 3.

To test for 8: A number is divisible by 8 if and only if the number formed by its last three digits is divisible by 8. (This follows from the fact that all multiples of 1,000 are divisible by 8.) Otherwise the remainder is the same as the remainder when the original number is divided by 8. (This rule holds for all powers of 2. A number is divisible by 2^n if and only if the last n digits form a number divisible by 2^n.)

To test for 9: A number is divisible by 9 if and only if it has a digital root of 9. If not, the digital root equals the remainder. The serial number of the bill has a digital root of 1, therefore it has a remainder of 1 when divided by 9.

To test for 10: A number is divisible by 10 if and only if it ends in 0. Otherwise the final digit equals the remainder.

To test for 11: Take the digits in a right to left order, alternately subtracting and adding. Only if the final result is divisible by 11 will the original number be divisible by 11. (It is assumed that 0 is divisible by 11.) Applied to the number on the bill, $2 - 4 + 1 - 1 + 7 - 6 + 1 - 6 = -6$. The final figure is not a multiple of 11, therefore neither is the original number. To determine the remainder, consider the final figure. If it is less than 11, and positive, it is the remainder. If it is negative, add 11 to find the remainder. If the final figure is more than 11, reduce it to a number less than 11 by dividing by 11 and putting down the excess. If the excess is positive, it is the remainder you seek; if it is negative, add 11. (In the example, $-6 + 11 = 5$. This tells you that the bill's number, divided by 11, has a remainder of 5.)

To test for 12: Test for 3 and 4, factors of 12. The number must meet both tests to be divisible by 12.

The reader has surely noticed a singular omission from the foregoing rules. How does one test for 7, the divine number of medieval numerology? It is the only digit for which no one has yet found a simple rule. This disorderly behavior on the part of 7 has long fascinated students of number theory. Dozens of curious 7 tests have been devised, all seemingly unrelated to one another; all, unfortunately, are almost as time-consuming as the orthodox division procedure.

One of the oldest of such tests is to take the digits of a number in reverse order, right to left, multiplying them successively by the digits, 1, 3, 2, 6, 4, 5, repeating with this sequence of multipliers as long as necessary. The products are added. The original number is divisible by 7 if and only if this sum is a multiple of 7. If the sum is not a multiple, its excess over a multiple of 7 equals the remainder when the original number is divided by 7. This is how the method is applied to the number on the bill:

$$
\begin{array}{rcr}
2 \times 1 &=& 2 \\
4 \times 3 &=& 12 \\
1 \times 2 &=& 2 \\
1 \times 6 &=& 6 \\
7 \times 4 &=& 28 \\
6 \times 5 &=& 30 \\
1 \times 1 &=& 1 \\
6 \times 3 &=& \underline{18} \\
&& 99
\end{array}
$$

Ninety-nine divided by 7 has an excess of 1. This is the remainder when the bill's number is divided by 7. The test can be speeded up by "casting out 7's" from the products: writing 5 instead of 12, 0 instead of 28 and so on. The sum will then be 22 instead of 99. The test is really nothing more than a method of casting multiples of 7 out of the original number. It derives from the fact that successive powers of 10 are congruent (modulo 7) to digits in the repeating series 1, 3, 2, 6, 4, 5; 1, 3, 2, 6, 4, 5 . . . (Numbers are congruent modulo 7 if they have the same remainder when divided by 7.) Instead of 6, 4, 5 one can substitute the congruent (modulo 7) multi-

pliers −1, −3, −2. The interested reader will find it all clearly
explained in the chapter on number congruence in *What
Is Mathematics?* by Richard Courant and Herbert Robbins
(1941). Once the basic idea is understood it is easy to invent
similar tests for any number whatever, as Blaise Pascal explained back in 1654. For example, to test for 13 we have
only to note that the powers of 10 are congruent (modulo
13) to the repeating series 1, −3, −4, −1, 3, 4 . . . This series
is applied to a number in the same manner as the series in
the test for 7.

What series of multipliers results when we apply this
method to divisibility by 3, 9 and 11? The powers of 10 are
congruent (modulo 3 and modulo 9) to the series 1, 1, 1, 1 . . . ,
so we arrive at once at the previously stated rules for 3 and
9. The powers of 10 are congruent (modulo 11) to the series
− 1, + 1, −1, +1 . . . , which leads to the previously stated rule
for 11. The reader may enjoy finding the multiplier series
for the other divisors to see how each series links up with
its corresponding rule, or in the cases of 6 and 12, leads to
other rules.

A bizarre 7 test, attributed to D. S. Spence, appeared in
1956 in *The Mathematical Gazette* (October, page 215). (The
method goes back to 1861; see L. E. Dickson, *History of The
Theory of Numbers*, Vol. 1, page 339, where it is credited to
A. Zbikovski of Russia.) Remove the last digit, double it,
subtract it from the truncated original number and continue
doing this until one digit remains. The original number is
divisible by 7 if and only if the final digit is 0 or 7. This
procedure is applied to our serial number in this manner:

$$
\begin{array}{r}
6167114\cancel{2} \\
4 \\
\hline
616711\cancel{0} \\
0 \\
\hline
61671\cancel{1} \\
2 \\
\hline
6166\cancel{9} \\
18 \\
\hline
614\cancel{8} \\
16 \\
\hline
59\cancel{8} \\
16 \\
\hline
4\cancel{3} \\
6 \\
\hline
-2
\end{array}
$$

The final digit is not divisible by 7, therefore neither is the original number. A defect of the system is that it gives no simple clue to the remainder.

The 7 test that seems to me the most efficient, especially when applied to very large numbers, is one developed by L. Vosburgh Lyons, a New York neuropsychiatrist. It is disclosed here for the first time in Figure 72, where the steps are applied to an arbitrary thirteen-digit number. The method is extremely rapid when applied to a six-digit number; one has only to build a triangle of three digits, then two, and then a final digit that provides the remainder.

Working with this method, Lyons has discovered many remarkable six-digit-number feats of the "lightning calculator" type. Here is one that appeared in *Ibidem*, No. 5, April 1956.

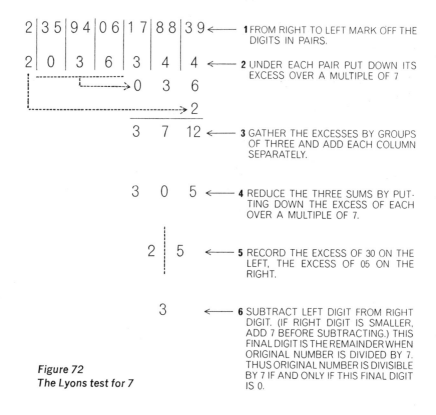

Figure 72
The Lyons test for 7

Ask someone to chalk on a blackboard any six-digit number that is *not* divisible by 7. Suppose he writes 431576. You propose to alter quickly each digit in turn, forming six new numbers, each a multiple of 7.

To do this, first write the number six times in a square array (as shown at the left in Figure 73), leaving a blank

4	3	1	5	7	**A**
4	3	1	5	**B**	6
4	3	1	**C**	7	6
4	3	**D**	5	7	6
4	**E**	1	5	7	6
F	3	1	5	7	6

4	3	1	5	7	1
4	3	1	5	3	6
4	3	1	6	7	6
4	3	6	5	7	6
4	7	1	5	7	6
3	3	1	5	7	6

Figure 73
A calculation stunt involving divisibility by 7

space for the last digit in the first row, the next to the last digit in the second row and so on. (The spaces are labeled A to F only to help the explanation; when the trick is performed, the six spaces are left blank.) Having already tested the number to make sure it is not divisible by 7, you have determined that is has an excess of 5. Obviously 1 must be placed in space A instead of the original 6 to make the top number a multiple of 7.

The remaining five blanks can now be filled in rapidly. In row 2, consider the number B6. Above it is 71, which has an excess of 1, when divided by 7. You must therefore place a digit in space B so that the number B6 will also have an excess of 1. This is done by placing a 3 in space B. (In your mind, simply subtract 1 from 6 to get 5, then ask yourself what two-digit multiple of 7 ends in 5. The answer can only be 35.) The number C7 is handled in the same way. Above it is 53, which has an excess of 4, so to give C7 a similar excess you put 6 in space C. Continue in similar fashion with the remaining rows. The final result is shown at the right in the illustration. Each row is now divisible by 7. To a mathematician familiar with the difficulties of testing divisibility by 7, the feat is quite astonishing. The trick is easy to do, of course, if the numbers sought are to be divisible by 9.

A knowledge of divisibility rules often furnishes short cuts in solving number problems that otherwise would be enormously difficult. For instance, if nine playing cards, with values from ace to nine are arranged at random to form a nine-digit number, what is the probability that it will be divisible by 9? Since the sum of the digits from 1 to 9 is 45, which has a digital root of 9, you know at once that the probability is 1 (certainty). Four cards, from ace to four, are randomly arranged. What is the probability that this four-digit number is divisible by 3? Bearing in mind the rule for 3, you know immediately that the probability is 0 (impossible).

A pleasant parlor trick begins by handing someone nine playing cards with values from ace to nine. While your back is turned, ask someone to arrange the ace, two, three and four in any order they please to make a four-digit number. Without turning around, you can tell them that the number is not divisible by 3. Now ask them to add the five and rearrange the cards to make a five-digit number. You assure them, back still turned, that the new number *is* divisible by 3.

Before looking at the answers the reader may enjoy testing his skill on the following digital puzzles, all intimately related to this chapter's topic.

1. A person older than nine and younger than a hundred is asked to write his age three times to make a six-digit number (e.g., 484848). Prove that the number must be divisible by 7.

2. Seven different playing cards, with values from ace to seven, are shaken in a hat, then taken out singly and placed in a row. What is the probability that this seven-digit number is divisible by 11?

3. Find the smallest number that has a remainder of 1 when divided by 2, a remainder of 2 when divided by 3, a remainder of 3 when divided by 4, a remainder of 4 when divided by 5, a remainder of 5 when divided by 6, a remainder of 6 when divided by 7, a remainder of 7 when divided by 8, a remainder of 8 when divided by 9 and a remainder of 9 when divided by 10.

4. A child has at his disposal n small wooden cubes, all the same size. With them he tries to build the largest cube he can, but discovers that he is short by exactly one row of small cubes that would have formed an edge of the large cube. Prove that n is divisible by 6.

5. What is the remainder when 3, raised to the power of 123,456,789, is divided by 7?

6. Find four different digits, excluding 0, which cannot be arranged to make a four-digit number divisible by 7.

The problems are easier than one might think at first, once they are approached properly, except for the last one, which seems to yield only to brute hammer-and-tongs methods. But any reader who solves all six will find that he has had a stimulating workout in elementary number theory.

ADDENDUM

My column on divisibility tests prompted a flood of letters. Many readers sent explanations of why the Zbikovski method works, how it can be applied to divisibility by other primes than 7, and procedures by which the remainder can be determined. Explanations of Lyons's method also came in, most of them too technical for me to understand.

Scores of readers provided other methods of testing for 7. I give here only the procedure mentioned by the largest number of correspondents. It is old and well known, deriving from the pleasant fact that 1,001 (the number, incidentally, of stories in the original *Arabian Nights*) is the product of the three consecutive primes: 7, 11 and 13. The number to be tested is partitioned into three-digit sections, starting at the right. For example, 61671142 is split into 61/671/142.

Alternately add and subtract these sections, starting at the right: $142 - 671 + 61 = -148$. The result has the same remainder when divided by 7, 11 or 13 as does the original number.

Readers interested in learning other divisibility tests for 7 and other numbers will find the first reference in the bibliography a handy check list of the older literature. I have added a list of recent articles that are readily accessible.

ANSWERS

1. To prove that a number of the form ABABAB must be evenly divisible by 7, we have only to note that such a number is the product of AB and 10101. Because 10101 is a multiple of 7, the number ABABAB must be also.

2. When the digits 1 to 7 are randomly arranged to form a number, the probability that the number is divisible by 11 is 4/35. To be divisible by 11 the digits must be arranged so that the difference between the sum of one set of alternate digits and the sum of the other set of alternate digits is either 0 or a multiple of 11. The sum of all seven digits is 28. It is easy to find that 28 can be partitioned in only two ways that meet the 11 test: 14|14 and 25|3. The 25|3 partition is ruled out because no sum of three different digits can be as low as 3. Therefore only the 14|14 partition need be considered. There are 35 different combinations of three digits that can fall into the B positions in the number ABABABA. Of the 35, only four (167, 257, 347, 356) sum to 14. Therefore the probablity that the number will be divisible by 11 is 4/35.

3. The smallest number that has a remainder of one less than the divisor, when divided by each integer from 2 to 10 inclusive, is 2519. It is amusing to note that "Professor Hoffmann," in his book *Puzzles Old and New* (1893) calls this a "difficult problem" and devotes more than two pages to solving it by a complicated application of divisibility rules. Hoffmann failed to note that each division falls just one short of

being exact, so we need only to find the lowest common multiple of 2, 3, 4, 5, 6, 7, 8, 9, 10, which is 2520, then subtract 1 to get the answer.

4. The problem of the cube with the missing edge of smaller cubes is equivalent to showing that a number of the form $n^3 - n$ (where n is any integer) must always be evenly divisible by 6. The following is perhaps the simplest proof:

$$n^3 - n = n \ (n^2 - 1) = n \ (n - 1) \ (n + 1).$$

The expression to the right of the second equal sign reveals that the number $(n^3 - n)$ is the product of three consecutive integers. In any set of three consecutive integers, it is easy to see that one integer must be divisible exactly by 3 and that at least one integer must be even. (These two properties may, to be sure, unite in the *same* integer, e.g., 17, *18*, 19.) Since 2 and 3 are factors of the product of the three consecutive integers, the product must be divisible by 2×3, or 6.

5. The remainder, when 3 to the power of 123456789 is divided by 7, is 6. The short cut here is that successive powers of 3, when divided by 7, have remainders that repeat endlessly the six-digit cycle 3, 2, 6, 4, 5, 1. Divide 123456789 by 6 to obtain a remainder of 3, then note the third digit in the cycle. It is 6, the answer to the problem.

Any number raised to successive powers and divided by 7 has remainders that repeat a cyclic pattern, and the pattern is the same for all numbers that are equivalent modulo 7. Any power of a number that is 1 (mod 7) has a remainder of 1 when divided by 7. Powers of numbers that are 2 (mod 7) have the remainder cycle: 2, 4, 1. The cycle for powers of numbers that are 3 (mod 7) is given above; for 4 it is 4, 2, 1; for 5 it is 5, 4, 6, 2, 3, 1; for 6 it is 6, 1; and for 7 it is, of course, 0.

What is the remainder when 123456789 is raised to the power of 123456789 and divided by 7? Since 123456789 is 1 (mod 7), we know at once that the remainder is 1.

6. The problem asked for a set of four different digits, excluding 0, that could not be arranged to make a four-digit number divisible by 7. Of the 126 different combinations of four digits, only three work: 1238, 1389 and 2469.

CHAPTER FIFTEEN

O

Nine Problems

1. The Seven File Cards

A SHEET of legal-sized paper, 8½ by 12½ inches, has an area of 106¼ square inches. Seven file cards of the three-by-five-inch size have a combined area of 105 square inches. Obviously it is not possible to cover the large sheet completely with the seven cards, but what is the largest area that *can* be covered? The cards must be placed flat, and they may not be folded or cut in any way. They may overlap the edges of the sheet, however, and it is not necessary for their sides to be parallel with the sides of the sheet. Figure 74 shows how the seven cards can be arranged to cover an area of 98¾ square inches. This is not the maximum.

Everyone in the family, young and old, will enjoy working on this puzzle. If the required materials are not handy, a sheet of cardboard can be cut to the 8½-by-12½-inch size, and the seven three-by-five rectangles can be cut from paper. It is a good plan to rule the large sheet into half-inch squares so that the area left exposed can be computed quickly.

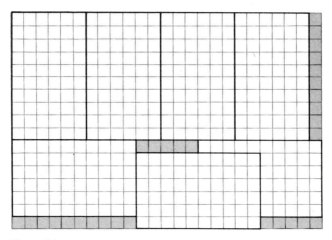

Figure 74
How much of the paper can be covered with seven file cards?

The problem was first posed by Jack Halliburton in *Recreational Mathematics Magazine,* December 1961.

2. A Blue-Empty Graph

SIX HOLLYWOOD STARS form a social group that has very special characteristics. Every two stars in the group either mutually love each other or mutually hate each other. There is no set of three individuals who mutually love one another. Prove that there is at least one set of three individuals who mutually hate each other. The problem leads into a fascinating new field of graph theory, "blue-empty chromatic graphs," the nature of which will be explained when the answer is given.

3. Two Games in a Row

A CERTAIN MATHEMATICIAN, his wife and their teen-age son all play a fair game of chess. One day when the son asked his father for ten dollars for a Saturday-night date, his father puffed his pipe a moment and replied:

"Let's do it this way. Today is Wednesday. You will play a game of chess tonight, another tomorrow and a third on Friday. Your mother and I will alternate as opponents. If you win two games in a row, you get the money."

"Whom do I play first, you or Mom?"

"You may have your choice," said the mathematician, his eyes twinkling.

The son knew that his father played a stronger game than his mother. To maximize his chance of winning two games in succession, should he play father-mother-father or mother-father-mother?

Leo Moser, a mathematician at the University of Alberta, is responsible for this amusing question in elementary probability theory. Of course you must prove your answer, not just guess.

4. A Pair of Cryptarithms

IN MOST CRYPTARITHMS a different letter is substituted for each digit in a simple arithmetical problem. The two remarkable cryptarithms shown in Figure 75 are unorthodox in their departure from this practice, but each is easily solved by logical reasoning and each has a unique answer.

```
    E E O              P P P
      O O                P P
    ───────            ───────
  E O E O            P P P P
  E O O            P P P P
  ─────────        ─────────
O O O O O          P P P P P
```

Figure 75
Two unorthodox cryptarithms

In the multiplication problem at the left in the illustration, newly devised by Fitch Cheney of the University of Hartford, each E stands for an even digit, each O for an odd digit. The fact that every even digit is represented by E does not mean, of course, that all the even digits are the same. For example, one E may stand for 2, another for 4, and so on. Zero is considered an even digit. The reader is asked to reconstruct the numerical problem.

In the multiplication problem at the right, each P stands for a prime digit (2, 3, 5, or 7). This charming problem was first proposed some twenty-five years ago by Joseph Ellis Trevor, a chemist at Cornell University. It has since become a classic of its kind.

5. Dissecting a Square

IF ONE FOURTH of a square is taken from its corner, is it possible to dissect the remaining area into four congruent (same size and shape) parts? Yes, it can be done in the manner shown at the left in Figure 76. Similarly, an equilateral triangle with one fourth of its area cut from a corner, as in the center figure in the illustration, can also be divided into four congruent parts. These are typical of a large variety of geometric puzzles. Given a certain geometric figure, the task is to cut it into a specified number of identical shapes that completely fill the larger figure.

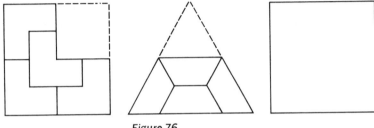

Figure 76
Three dissection puzzles

Can the square at the right in the illustration be dissected into five congruent parts? Yes, and the answer is unique. The pieces can be any shape, however complex or bizarre, provided that they are identical in shape and size. An asymmetric piece may be "turned over"; that is, it is considered identical with its mirror image. The problem is annoyingly intractable until suddenly the solution strikes like lightning.

6. Traffic Flow in Floyd's Knob

ROBERT ABBOTT, author of *Abbott's New Card Games* (New York: Stein and Day, 1963), provided the curious street map reproduced in Figure 77, accompanied by the following story:

"Because the town of Floyd's Knob, Indiana, had only thirty-seven registered automobiles, the mayor thought it would be safe to appoint his cousin, Henry Stables, who was the town cutup, as its traffic commissioner. But he soon regretted his decision. When the town awoke one morning, it found that a profusion of signs had been erected establishing numerous one-way streets and confusing restrictions on turns.

"The citizens were all for tearing down these signs until the police chief, another cousin of the mayor, made a surprising discovery. Motorists passing through town became so exasperated that sooner or later they made a prohibited turn. The police chief found that the town was making even more money from these violations than from its speed trap on an outlying country road.

"Of course everyone was overjoyed, particularly because the next day was Saturday and Moses MacAdam, the county's richest farmer, was due to pass through town on his way to the county seat. They expected to extract a large fine from Moses, believing it to be impossible to drive through town without at least one traffic violation. But Moses had been secretly studying the signs. When Saturday morning came, he astonished the entire town by driving from his farm through town to the county seat without a single violation!

"Can you discover the route Moses took? At each intersec-

Figure 77
The traffic maze in Floyd's Knob

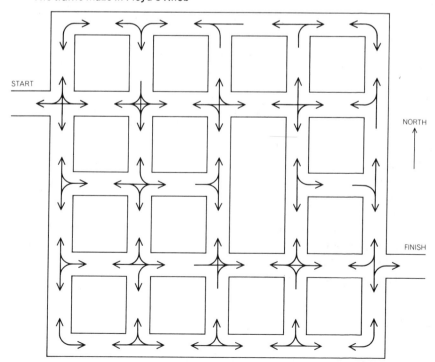

tion you must follow one of the arrows. That is, you may turn in a given direction only when there is a curved line in that direction, and you may go straight only when there is a straight line to follow. No turns may be made by backing the car around a corner. No U-turns are permitted. You may leave an intersection only at the head of an arrow. For instance, at the first intersection after leaving the farm, you have only two choices: to go north or to go straight. If you go straight, at the next intersection you must either go straight or turn south. True, there is a curved line to the north, but there is no arrow pointing north, so you are forbidden to leave that intersection in a northerly direction."

7. Littlewood's Footnotes

EVERY NOW AND THEN a magazine runs a cover picture that contains a picture of the same magazine, on the cover of which one can see a still smaller picture of the magazine, and so on presumably to infinity. Infinite regresses of this sort are a common source of confusion in logic and semantics. Sometimes the endless hierarchy can be avoided, sometimes not. The English mathematician J. E. Littlewood, commenting on this topic in his *A Mathematician's Miscellany* (London: Methuen, 1953), recalls three footnotes that appeared at the end of one of his papers. The paper had been published in a French journal. The notes, all in French, read:

"1. I am greatly indebted to Prof. Riesz for translating the present paper.

"2. I am indebted to Prof. Riesz for translating the preceding footnote.

"3. I am indebted to Prof. Riesz for translating the preceding footnote."

Assuming that Littlewood was completely ignorant of the French language, on what reasonable grounds did he avoid an infinite regress of identical footnotes by stopping after the third footnote?

8. Nine to One Equals 100

AN OLD NUMERICAL PROBLEM that keeps reappearing in puzzle books as though it had never been analyzed before is the problem of inserting mathematical signs wherever one likes between the digits 1, 2, 3, 4, 5, 6, 7, 8, 9 to make the expression equal 100. The digits must remain in the same sequence. There are many hundreds of solutions, the easiest to find perhaps being

$$1 + 2 + 3 + 4 + 5 + 6 + 7 + (8 \times 9) = 100.$$

The problem becomes more of a challenge if the mathematical signs are limited to plus and minus. Here again there are many solutions, for example

$$1 + 2 + 34 - 5 + 67 - 8 + 9 = 100,$$
$$12 + 3 - 4 + 5 + 67 + 8 + 9 = 100,$$
$$123 - 4 - 5 - 6 - 7 + 8 - 9 = 100,$$
$$123 + 4 - 5 + 67 - 89 = 100,$$
$$123 + 45 - 67 + 8 - 9 = 100,$$
$$123 - 45 - 67 + 89 = 100.$$

"The last solution is singularly simple," writes the English puzzlist Henry Ernest Dudeney in the answer to Problem No. 94 in his *Amusements in Mathematics,* "and I do not think it will ever be beaten."

In view of the popularity of this problem it is surprising that so little effort seems to have been spent on the problem in reverse form. That is, take the digits in descending order, 9 through 1, and form an expression equal to 100 by inserting the smallest possible number of plus or minus signs.

9. The Crossed Cylinders

ONE OF ARCHIMEDES' greatest achievements was his anticipation of some of the fundamental ideas of calculus. The problem illustrated in Figure 78 is a classic example of a problem that most mathematicians today would regard as

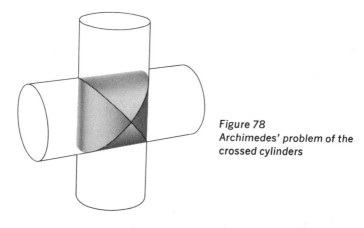

Figure 78
Archimedes' problem of the
crossed cylinders

unsolvable without a knowledge of calculus (indeed, it is found in many calculus textbooks) but that yielded readily to Archimedes' ingenious methods. The two circular cylinders intersect at right angles. If each cylinder has a radius of one unit, what is the volume of the shaded solid figure that is common to both cylinders?

No surviving record shows exactly how Archimedes solved this problem. There is, however, a startlingly simple way to obtain the answer; indeed, one need know little more than the formula for the area of a circle (pi times the square of the radius) and the formula for the volume of a sphere (four-thirds pi times the cube of the radius). It may have been the method Archimedes used. In any case, it has become a famous illustration of how calculus often can be side-stepped by finding a simple approach to a problem.

ANSWERS

1

If it is required that the cards be placed with their edges parallel to the edges of the sheet, a maximum of 100 square inches can be covered. Figure 79 shows one of many different ways in which the cards can be placed.

Figure 79
Seven file cards arranged to cover 100 square inches

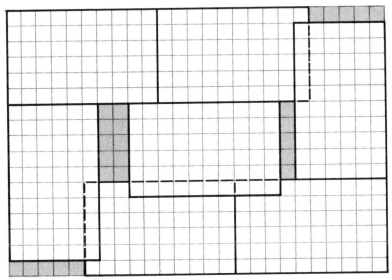

Stephen Barr was the first to point out that if the central card is tilted as shown in Figure 80 the covered area can be increased to 100.059+ square inches. Then Donald Vanderpool wrote to say that if the card is rotated a bit more than this (keeping it centered on the exposed strip) the covered area can be increased even more. The angle at which maximum coverage is achieved must be obtained by calculus.

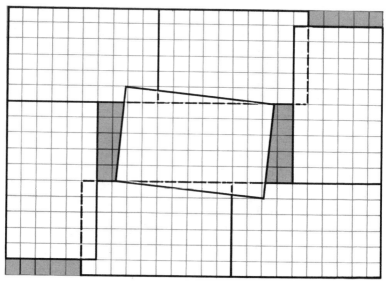

Figure 80
The file cards arranged to cover a fraction of an inch more

James A. Block found that the angle could be varied from 6° 12′ to 6° 13′ without altering the covered area expressed to five decimal places: 100.06583+ square inches. The problem's history has been sketched in Joseph S. Madachy, *Mathematics on Vacation* (New York: Scribner's, 1966), pages 133–35. He gives a calculation by R. Robinson Rowe for an angle of 6° 12′ 37.8973″ that gives a covering of 100.065834498+ square inches.

2

Every two people in a set of six people either mutually love or mutually hate each other, and there is no set of three who mutually love one another. The problem is to prove that there is a set of three who mutually hate one another.

The problem is easily solved by a graph technique. Six dots represent the six individuals (*see Figure 81*). All possible pairs are connected by a broken line that stands for either mutual love or mutual hate. Let blue lines symbolize love and red lines symbolize hate.

Consider dot *A*. Of the five lines radiating from it, at least three must be of the same color. The argument is the same regardless of which color or which three lines we pick, so let us assume three lines are red (shown black in the illustration). If the lines forming triangle *BCE* are all blue, then we have a set of three people who mutually love one another. We are told no such set exists; therefore at least one side of this triangle must be red. No matter which side we pick for red, we are sure to form an all-red triangle (i.e., three people who mutually hate one another). The same result is obtained if we choose to make the first three lines blue instead of red. In that case the sides of triangle *BCE* must all be red; otherwise a blue side would form an all-blue triangle. In brief, there must be at least one triangle that is either all-blue or all-red. The problem rules out an all-blue triangle, so there must be an all-red one.

Actually, a stronger conclusion is obtainable. If there is no all-blue triangle, it can be shown (by more complicated reasoning) that there are at least *two* all-red triangles. In graph theory, a two-color graph of this sort, with no blue triangles, is called a blue-empty chromatic graph. If the number of points is six, as in this problem, the minimum number of red triangles is two.

When the number of points in a blue-empty graph is less

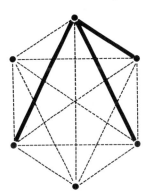

Figure 81
Graph solution for Problem 2

than six, it is easy to draw such graphs with no red triangles. When the number of points is seven, there must be at least four red triangles. For an eight-point blue-empty graph the minimum number of red triangles is eight; for a nine-point graph it is thirteen. Anyone wishing to go deeper into the topic can consult the following papers:

R. E. Greenwood and A. M. Gleason, "Combinatorial Relations and Chromatic Graphs," *Canadian Journal of Mathematics*, Vol. 7 (1955), pages 1–7.

Leopold Sauve, "On Chromatic Graphs," *American Mathematical Monthly*, Vol. 68, February 1961, pages 107–11.

Gary Lorden, "Blue-empty Chromatic Graphs," *American Mathematical Monthly*, Vol. 69, February 1962, pages 114–20.

J. W. Moon and Leo Moser, "On Chromatic Bipartite Graphs," *Mathematics Magazine*, Vol. 35, September 1962, pages 225–27.

J. W. Moon, "Disjoint Triangles in Chromatic Graphs," *Mathematics Magazine*, Vol. 39, November 1966, pages 259–61.

3

A plays a stronger chess game than *B*. If your object is to win two games in a row, which is better: to play against *A*, then *B*, then *A;* or to play *B*, then *A*, then *B?*

Let P_1 be the probability of your defeating *A* and P_2 the probability of your defeating *B*. The probability of your *not* winning against *A* will then be $1 - P_1$ and the probability of your not winning against *B* will be $1 - P_2$.

If you play your opponents in the order *ABA*, there are three different ways you can win two games in a row:

1. You can win all three games. The probability of this occurring is $P_1 \times P_2 \times P_1 = P_1^2 P_2$.

2. You can win the first two games only. The probability of this is $P_1 \times P_2 \times (1 - P_1) = P_1 P_2 - P_1^2 P_2$.

3. You can win the last two games only. The probability is $(1 - P_1) \times P_2 \times P_1 = P_1 P_2 - P_1^2 P_2$.

The three probabilities are now added to obtain $P_1 P_2 (2 - P_1)$. This is the probability that you will win twice in a row if you play in the order *ABA*.

If the order is *BAB*, a similar calculation will show that the probability of winning all three games is $P_1P_2^2$, of winning the first two games is $P_1P_2 - P_1P_2^2$, and of winning the last two games is $P_1P_2 - P_1P_2^2$. The sum of the three probabilities is $P_1P_2(2 - P_2)$. This is the probability of winning two games in a row if you play in the order *BAB*.

We know that P_2, which is the probability of your winning against *B*, is greater than P_1, the probability of your winning against *A*, so it is apparent that P_1P_2 $(2 - P_1)$ must be greater than P_1P_2 $(2 - P_2)$. In other words, you stand a better chance of winning twice in succession if you play *ABA*: first the stronger player, then the weaker, then the stronger.

Fred Galvin, Donald MacIver, Akiva Skidell, Ernest W. Stix, Jr., and George P. Yost were the first of many readers who reached the same conclusion by the following informal reasoning. To win two games in a row it is essential that the son win the second game, therefore it is to his advantage to play the second game against the weaker player. In addition, he must win at least once against the stronger player, therefore it is to his advantage to play the stronger player twice. Ergo, *ABA*. Galvin pointed out that if the problem can be solved without knowing the probabilities involved, the answer will be obtainable from any special case. Consider the extreme case when the son is certain to beat his mother. He then is sure to win two games in a row if he can beat his father once, so obviously he increases his chances by playing his father twice.

4

Fitch Cheney's cryptarithm has the unique answer

$$
\begin{array}{r}
285 \\
39 \\
\hline
2565 \\
855 \\
\hline
11115
\end{array}
$$

The unique answer to Joseph Ellis Trevor's cryptarithm is

$$\begin{array}{r} 775 \\ 33 \\ \hline 2325 \\ 2325 \\ \hline 25575. \end{array}$$

Trevor's problem, the more difficult of the two, is perhaps best approached by searching first for all three-digit numbers composed of prime digits that yield four prime digits when multiplied by a prime. There are only four:

$$775 \times 3 = 2325,$$
$$555 \times 5 = 2775,$$
$$755 \times 5 = 3775,$$
$$325 \times 7 = 2275.$$

No three-digit number has more than one multiplier, therefore the multiplier in the problem must consist of two identical digits. Thus there are only four possibilities that need to be tested.

5

A square can be dissected into five congruent parts only in the manner shown in Figure 82. The consternation of those who find themselves unable to solve this problem is equaled only by their feeling of foolishness when shown the answer.

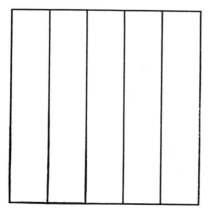

Figure 82
Solution for Problem 5

6

To drive through Floyd's Knob without a traffic violation, take the following directions at each successive intersection (the letters stand for North, South, East, West) : E-E-S-S-E-N-N-N-E-S-W-S-E-S-S-W-W-W-W-N-N-E-S-W -S-E- E- E-E-N-E.

7

"However little French I know," says J. E. Littlewood (in explaining why he was not obliged to write an infinite regress of footnotes to an article that a friend translated), "I am capable of *copying* a French sentence."

8

To form an expression equal to 100, four plus and minus signs can be inserted between the digits, taken in reverse order, as follows :

$$98 - 76 + 54 + 3 + 21 = 100.$$

There is no other solution with as few as four signs. For a complete tabulation of all solutions for both the ascending and descending sequence see my *Numerology of Dr. Matrix* (New York: Simon and Schuster, 1967), pages 64–65.

9

Two circular cylinders of unit radius intersect at right angles. What is the volume common to both cylinders? The problem is solved easily, without the use of calculus, by the following elegant method:

Imagine a sphere of unit radius inside the volume common to the two cylinders and having as its center the point where the axes of the cylinders intersect. Suppose that the cylinders and sphere are sliced in half by a plane through the sphere's center and both axes of the cylinders *(at left of Figure 83)*. The cross section of the volume common to the cylinders will be a square. The cross section of the sphere will be a circle that fills the square.

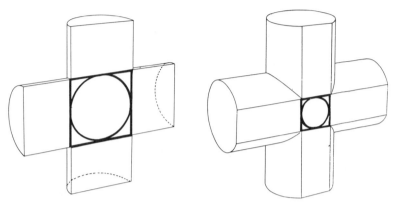

Figure 83
Two cross sections of Archimedes' cylinders and the interior sphere

Now suppose that the cylinders and sphere are sliced by a plane that is parallel to the previous one but that shaves off only a small portion of each cylinder (*at right of the illustration*). This will produce parallel tracks on each cylinder, which intersect as before to form a square cross section of the volume common to both cylinders. Also as before, the cross section of the sphere will be a circle inside the square. It is not hard to see (with a little imagination and pencil doodling) that any plane section through the cylinders, parallel to the cylinders' axes, will always have the same result: a square cross section of the volume common to the cylinders, enclosing a circular cross section of the sphere.

Think of all these plane sections as being packed together like the leaves of a book. Clearly, the volume of the sphere will be the sum of all circular cross sections, and the volume of the solid common to both cylinders will be the sum of all the square cross sections. We conclude, therefore, that the ratio of the volume of the sphere to the volume of the solid common to the cylinders is the same as the ratio of the area of a circle to the area of a circumscribed square. A brief calculation shows that the latter ratio is $\pi/4$. This allows the following equation, in which x is the volume we seek:

$$\frac{4\pi r^3/3}{x} = \frac{\pi}{4}$$

The π's drop out, giving x a value of $16r^3/3$. The radius in this case is 1, so the volume common to both cylinders is 16/3. As Archimedes pointed out, it is exactly 2/3 the volume of a cube that encloses the sphere; that is, a cube with an edge equal to the diameter of each cylinder.

A number of readers pointed out that this solution makes use of what is called "Cavalieri's theorem," after Bonaventura Cavalieri, a seventeenth-century Italian mathematician. "This theorem in its simplest form," wrote Fremont Reizman, "says that two solids are equal in volume if they have equal altitudes and equal cross sections at equal heights above the base. But to prove it, Cavalieri had to anticipate calculus a bit by building his figures up from a stack of laminae and passing to the limit." The principle was known to Archimedes. In a lost book called *The Method* that was not found until 1906 (it is the book in which Archimedes gives the answer to the crossed-cylinders problem), he attributes the principle to Democritus, who used it for obtaining the formula for the volume of a pyramid or cone.

Several readers solved the problem by applying Cavalieri's theorem in a slightly different way. Granville Perkins, for instance, did it by circumscribing (around the solid common to both cylinders) a cube. Using the two faces parallel to both axes as bases, he constructed two pyramids with apices at the cube's center. By slicing laminae parallel to these bases the problem is easily solved.

For readers who care to tackle the problem of finding the volume of the solid common to *three* orthogonally intersecting cylinders of unit radius, I give only the solution: $8 (2 - \sqrt{2})$. A calculus solution is given in S. I. Jones, *Mathematical Nuts* (privately printed in Tennessee, 1932), pages 83 and 287. Discussions of the two-cylinder case will be found in J. H. Butchart and Leo Moser, "No Calculus Please," *Scripta Mathematica*, Vol. 18, September 1952, pages 221–36, and Richard M. Sutton, "The 'Steinmetz Problem' and School Arithmetic," *Mathematics Teacher*, Vol. 50, October 1957, pages 434–35.

○

CHAPTER SIXTEEN

The Eight Queens and Other Chessboard Diversions

Pennypacker's office still smelled of linoleum, a clean, sad scent that seemed to lift from the checkerboard floor in squares of alternating intensity; this pattern had given Clyde as a boy a funny nervous feeling of intersection, and now he stood crisscrossed by a double sense of himself ...

—JOHN UPDIKE, *Pigeon Feathers*

THE CRISSCROSSING of a checked pattern may give some people a "nervous feeling," but when a recreational mathematician sees a checkerboard floor his mind leaps happily toward puzzle possibilities. It is safe to say that no other geometrical pattern has been so thoroughly exploited for recreational purposes. I am not referring now to games such as checkers, chess and go, which use the checked pattern as a board, but to an endless variety of puzzles that derive from the metric and topological properties of the pattern itself.

Consider for a moment a problem that appeared in my column in 1957 and is now well known. If two diagonally opposite corner squares of an 8 × 8 checkerboard are removed, can the remaining 62 cells be completely covered by 31 dominoes? Since each domino is assumed to cover two adjacent squares, one black and one white, 31 dominoes must cover 31 black squares and 31 white squares. But diagonally opposite corner squares are the same color, so the mutilated board will have 32 squares of one color and 30 of another and clearly cannot be covered by 31 dominoes. This proof of impossibility is a classic illustration of how the coloring of a checkerboard, far from merely making the pattern more pleasing aesthetically or more convenient for plotting checker and chess moves, provides a powerful tool for analyzing many types of checkerboard problems.

Instead of removing two squares of the same color, suppose we remove two squares of opposite colors. They may be taken from any two spots on the board. Is it always possible to cover the remaining 62 squares with 31 dominoes? The answer is yes. But is there a simple way to prove it? One could, of course, test all possible combinations of missing squares, but that would be tedious and inelegant. Dana Scott, a mathematician at the University of California, has called to my attention a beautiful proof discovered by his friend Ralph Gomory, a research mathematician. Heavy lines are drawn on the board as shown in Figure 84, forming a closed path along which the cells lie like beads of alternating colors on a necklace. The removal of two squares of opposite colors from any two spots along this path will cut the path into two open-ended segments (or one segment if the removed squares are adjacent on the path). Since each segment must consist of an even number of squares, each segment (and therefore the entire board) can be completely covered by dominoes.

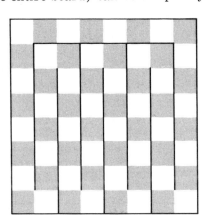

Figure 84
Gomory's proof of a domino-
and-checkerboard theorem

Instead of trying to cover a mutilated checkerboard with dominoes, suppose we mutilate it in such a way that *no* domino can be placed on it. What is the smallest number of squares that must be taken away in order to make it impossible to place a single domino on what remains? It is easy to see that 32 squares, all of one color, must be removed. But the problem is not so easy to solve if we substitute for the domino one of the higher "polyominoes." (A polyomino is any figure formed by checkerboard squares that are connected along their edges.) Solomon W. Golomb, a mathematician at the University of Southern California, and author of *Polyominoes* (New York: Scribner's, 1965), has recently proposed this type of problem and answered it for every type of polyomino up through the twelve pentominoes (five-square figures). The pentomino shaped like a Greek cross provides a pretty problem. Assume that the 8 × 8 checkerboard is made of paper. If sixteen squares are shaded as shown in Figure 85, it obviously is not possible to cut a Greek cross from the unshaded squares. But sixteen is not the minimum. What *is* the minimum?

A fascinating checkerboard-cutting problem, as yet unsolved, is that of determining the number of different ways the 8 × 8 board can be cut in half along the solid lines that form the cells. The two halves must be the same size and shape so that one can be fitted on top of the other without flopping either one over. Henry Ernest Dudeney, the English puzzlist, first posed this problem and reported that he found it "bristling with difficulties." He was unable to make a full

Figure 85
Golomb's Greek-cross
problem

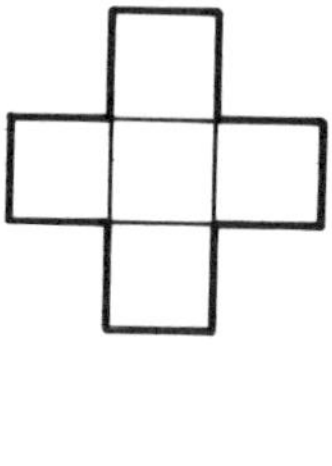

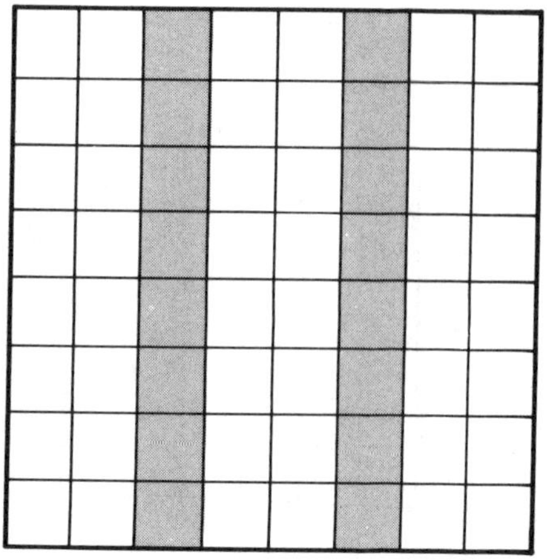

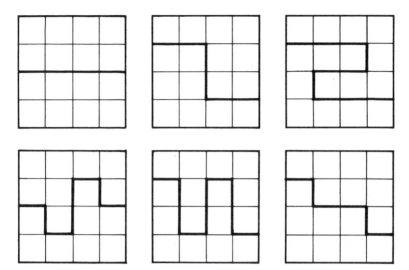

Figure 86
Six ways to halve a 4 X 4 board

tabulation of patterns. It is obvious that a 2 × 2 board can be cut in half in only one way. The 3 × 3 cannot be divided into identical parts (because it contains an odd number of cells), but if the central cell is considered a hole, it also can be bisected in only one way.

The 4 × 4 takes a bit of thinking, but it is not hard to discover that there are just six solutions (*see Figure 86*). These can be rotated and reflected in various ways, but patterns so obtained are not considered "different." Dudeney was able to show that the 5 × 5 (with missing center cell) has fifteen solutions and that the 6 × 6 has 255. There he stopped. The 7 × 7 and 8 × 8 problems should lend themselves easily to solution by a modern computer, but I am not aware that anyone has yet harnessed a computer for either task.

A closely related problem, first posed by Howard Grossman, a New York mathematics teacher, is that of cutting a square checkerboard into congruent quarters. As before, the four pieces must be the same size and shape and have the same "handedness." The coloring of the board is ignored. The 2 × 2 obviously can be quartered in only one way; the same is true of the 3 × 3 with the center hole. What about the 4 × 4? How many fundamentally different ways can it be quartered, not counting rotations and reflections? Readers should have little difficulty in drawing all the patterns. More ambitious readers may wish to go on to the 5 × 5 (with center hole), which has seven patterns. (The fact that it is possible to quarter any even-order board and any odd-order board with a center hole rests on the fact that the square of any even number is exactly divisible by 4 and the square of any odd number has a remainder of 1 when divided by 4.) Even the 6 × 6 is easily solved without computer aid, although the number of patterns rises to thirty-seven. As in the previous problem, solutions for the 7 × 7 and 8 × 8 are not known, unless somewhere a computer has devoted a few minutes during off hours to the contemplation of these problems.

Both the halving and the quartering of square checkerboards have their analogues in three dimensions, where the analysis is considerably more complex. Even the lowly 2 × 2 × 2 is tricky. Many people guess that there is but one way to halve such a cube (cutting only along planes that divide the cubical cells) when in fact there are three. (Can the reader visualize them?) It can be quartered in two ways. As for the 4 × 4 × 4, as far as I know, no one has the slightest notion of how many different ways it can be halved or quartered.

When counters of various sorts are added to the board, an infinite variety of puzzle possibilities open up. For example, given a checkerboard of order n (the order is the number of cells on a side), what is the largest number of chess queens that can be placed on the board in such a way that no queen is attacked by another? Since a queen moves an unlimited distance up and down, left and right and diagonally, the task is the same as that of placing a maximum number of counters so that no two lie in the same row, column or diagonal. It is easy to see that the maximum cannot exceed the order of the board, and it has been shown that on any board of order n, where n is greater than 3, n queens can be placed to meet the problem's conditions.

Not counting rotations and reflections as being different, there is only one way to place the queens on the 4 × 4 board, two ways on the 5 × 5, one way on the 6 × 6. (The reader may enjoy finding these patterns. The 6×6 problem has often been sold as a peg-and-board puzzle.) A 7×7 board has six solutions, the 8×8 has twelve, the 9×9 has forty-six, and the 10×10 has ninety-two. (There is no known formula by which the number of solutions on a board of order n can be determined.) When the order of the board is not divisible by 2 or 3, it is possible to superimpose n solutions that completely fill all the cells. Thus on the 5×5 one can place twenty-five queens—five of each of five colors—in such a way that no queen attacks another of the same color.

The twelve fundamental patterns for the standard 8×8 chessboard are shown in Figure 87. An enormous literature has grown up around this problem—usually called "the problem of the eight queens"—since it was first proposed by Max Bezzel in the Berlin *Schachzeitung,* September 1848, and the twelve solutions, by Franz Nauck, were published in 1850 in the Leipzig *Illustrierte Zeitung.* It is not easy to prove that the twelve patterns exhaust all possibilities. Such a proof, by way of determinants, was finally obtained by the English mathematician J. W. L. Glaisher and published in *Philosophical Magazine* for December 1874.

Each of the twelve basic solutions can be rotated and reflected to give seven other patterns, excepting Solution 10, which, because of its symmetry, yields only three other patterns. Thus there are ninety-two solutions altogether. Solution 10 is unique in having no queens on its sixteen central squares. It shares with Solution 1 the lack of queens along both main diagonals. Solution 7 is the most interesting of all: it is the only pattern in which no three queens (considered as points at the center of their cells) lie on a straight line. The reader may enjoy verifying this by finding straight lines on all the other patterns that pass through three or four queens. (The reference here is not to diagonals of squares on the board but to geometrical straight lines of any orientation.) Every now and then a puzzlist announces that he has found a second pattern that also avoids three-in-a-line, but on closer inspection it always turns out that there has been an oversight or that his second pattern is merely a rotation or reflection of Solution 7. Incidentally, it is sometimes maintained that the eight-queen problem has no solution with a queen on a corner cell; as the illustration shows, there are actually two such solutions. Note also that in every solution there must be a queen on a border cell that is the fourth from a corner.

Other chess pieces can, of course, be substituted for queens.

Figure 87
The twelve solutions to the classic problem of eight queens

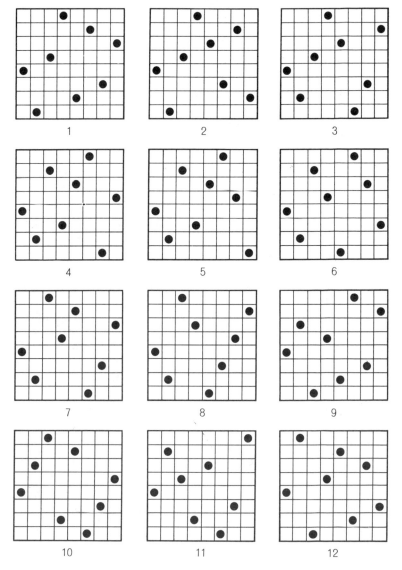

In the case of rooks it is obvious that, like the queens, a maximum of n rooks can be placed on a board of order n; more would put at least two rooks in one of the rows. A method that applies to a board of any size is simply to line the rooks up along a main diagonal. The number of ways this can be done is $n!$ (that is, $1 \times 2 \times 3 \ldots n$), but the task of eliminating rotation and reflection duplicates is so difficult that it is not known how many essentially different solutions exist even on as low-order a board as the 8×8.

For bishops the maximum is $2n - 2$. To prove this, note that the number of diagonals running in one direction is $2n - 1$. The two diagonals which consist of single squares, however, cannot both be occupied because two bishops would then be on a main diagonal running the other way. This reduces the maximum to $2n - 2$. Thus on the standard board no more than fourteen bishops can be placed so that no two attack each other. Dudeney has shown that this can be done in thirty-six essentially different ways. The total number of ways on a board of order n is 2^n, but (as with the rooks) it is not easy to winnow out the rotation and reflection duplicates. A method of placing the maximum number of bishops on a board of any size is to fill one edge row with n bishops and center $n - 2$ bishops along the opposite edge.

The maximum for kings is $n^2/4$ on even-order boards, $(n + 1)^2/4$ on odd-order boards. One pattern: the kings are arranged in a square lattice, each separated by one cell from all neighbors. The problem of determining the number of different ways of placing the maximum number of non-attacking kings on an $n \times m$ board is a difficult one. It was only recently solved, by Karl Fabel and C. E. Kemp. (See Eero Bonsdorff, Karl Fabel, and Olavi Riihimaa, *Schach und Zahl* [Düsseldorf: Walter Rau Verlag, 1966], pages 51–54.) Including rotations and reflections, there are 281,571 solutions on the 8×8 board.

The knight, which Dudeney calls the "irresponsible low comedian of the chessboard" because of its odd way of hopping, is perhaps less easy to analyze than the other pieces. What is the largest number of knights that can be placed on the 8×8 board in such a way that no knight attacks another? And in how many different ways can it be done?

ADDENDUM

The problem of halving and quartering square checker-boards caught the fancy of many readers. R. B. Tasker, Sherman Oaks, California, and William E. Patten, South Boston, Virginia, working independently and without computers, verified Dudeney's figure of 255 different ways of bisecting the order-6 board. John McCarthy, at the Stanford Computation Center, assigned to his students the problem of writing a computer program for orders 7 and 8. The results, which he sent to me in November 1962, are: 1,897 patterns for order-7, and 92,263 for order-8. As far as I know, this was the first determination of those figures. They were confirmed by later computer programs by Bruce Fowler, Pine Brook, New Jersey, and Norwood and Ruth Gove, Washington, D.C. Joh. Kraaijenhof, in Amsterdam, in 1963 sent the figure 1,972,653 for the order-9, and in 1966 Robert Maas, at the University of Santa Clara, reported 213,207,210 for the order-10. The order-9 result was confirmed in 1968 by Michael Cornelison of General Electric, Bethesda, Maryland, using a GE 635 GECOS system with a running time of 22 minutes.

I have not learned the details of the Stanford program. Fowler reports that his program is based on the fact that any bisecting line must pass through the board's center, and that the two halves of the line are symmetrical with respect to the center. "The program works somewhat like the mouse in a maze," he writes. "It starts at the center, moving one space at a time and making all possible right-hand turns. When it bumps into its previous path, it backs up one space, turns left 90 degrees and continues. When it reaches the edge of the board, it scores up one solution, backs up one space, turns left, and so on. In this way all possible solutions are obtained. The total is printed when it discovers the direct route to the edge, in the original starting direction."

The "backtrack" program just described applies only to the even-order boards. Odd-order boards, with center square removed, are more complicated.

On the quartering problem, the thirty-seven patterns for the order-6 board are given by Harry Langman in *Play Mathematics* (New York: Hafner, 1962), pages 127–28, and can be extracted from the pictures of the ninety-five patterns (including rotations and reflections) given in L. A. Graham, *Ingenious Mathematical Problems and Methods* (New York: Dover Publications, 1959), pages 164–65. John F. Moore, of Lockheed Electronics Corporation, Plainfield, New Jersey, was the first to determine (he did not use a computer) the 104 quartering patterns for the order-7 board, and the only reader to obtain the 766 patterns for the order-8. The order-7 result was independently obtained, without computer, by W. H. Grindley, Staffordshire, England, and by John Reed, Lexington, Massachusetts, who used a computer program written by Charles Peck and himself.

For readers interested in the eight-queens problem—its history, generalization, and curious sidelights—I have listed in the bibliography the best references I know. The Ahrens work is the fullest. The numbers of total and basic solutions for all orders through 13 are given by Ahrens. The number of total solutions is known for higher-order boards, but if the number of fundamental solutions for order-14 has been determined I have not yet learned of it.

Warren Lushbaugh, Los Angeles, called my attention to an elegantly simple proof that the twelve order-8 solutions cannot be superimposed to fill the sixty-four cells of the board. It is given by Thorold Gosset in *Messenger of Mathematics* (Vol. 44, July 1914, page 48). Sketch the 8×8 board then color the four middle cells of each edge and the four corner cells of the central 6×6 board. Inspection of the twelve ways the queens can be placed shows that each pattern has at least three queens on the twenty colored cells. If more than six solutions could be superimposed it would put at least twenty-one queens on the twenty colored cells, one queen to a cell, which is manifestly impossible.

An interesting variant of the queens problem is to give each queen the added power of a knight move. Can n nonattacking "superqueens" be placed on an order-n board? It is easy to prove that there are no solutions on boards through order-8. Nor is there one on order-9. Hilario Fernandez Long, Buenos Aires, examined the ninety-two patterns for the queens on the order-10 board and wrote that there is one pattern, and one only, that permits all ten queens to be superqueens and remain nonattacking. Readers may enjoy finding this unique pattern for themselves.

The problem of the nonattacking rooks on the standard chessboard was solved independently by two readers in 1962. David F. Smith, Cocoa Beach, Florida, and Donald B. Charnley, Los Angeles, each working without computers, found 5,282 basic solutions for the order-8 and 46,066 for the order-9 board. Charnley reported 456,454 for the order-10 board but this has not yet, to my knowledge, been confirmed. For readers who are interested, I have cited references on the rook problem in the bibliography. The number of fundamental solutions for orders 2 through 7 are, respectively, 1, 2, 7, 23, 115, 694.

ANSWERS

A minimum of ten squares must be removed from an 8 × 8 board to make it impossible to cut a five-square Greek cross from what remains. There are many solutions. The one shown in Figure 88 was provided by L. Vosburgh Lyons of New York City.

The 4 × 4 board can be quartered in no more than five different ways, shown at the top of Figure 89. Half of the second pattern can be reflected, but then two of the pieces will not have the same handedness as the other two. The seven ways of quartering the 5 × 5 (with center hole) are shown at the bottom of Figure 89.

A maximum of thirty-two knights can be placed on a standard chessboard in such a way that no knight attacks another. Simply place the knights on all squares of the same color. Jay Thompson of New York City writes that a group of chess players at a Middle Western hotel got into such a violent argument over this problem that the night clerk had to get a policeman to pull his chess nuts out of the foyer.

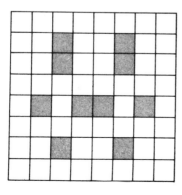

Figure 88
A solution to the Greek-cross problem

Figure 89

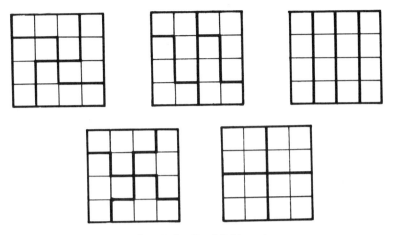

Quartering the 4 X 4 board

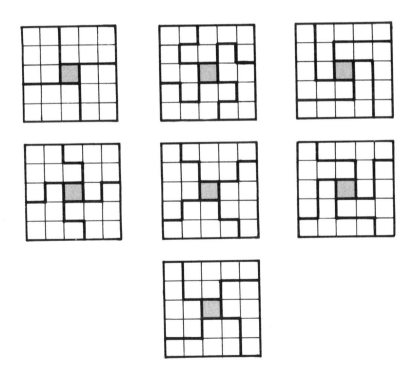

Quartering the 5 X 5 board

CHAPTER SEVENTEEN

○

A Loop of String

"JANE ELLIN JOYCE bubbled into our big new drugstore . . .
She hopped onto a stool at the soda counter and elbowed back
a black evening wrap from a low-cut white dress . . . She was
holding up her hands in front of her. There was a long loop
of cord between them."

So begins *Leopard Cat's Cradle,* an offbeat mystery novel
by Jerome Barry. An anthropologist at Columbia University
has initiated Jane Ellin into the mysteries of the string play
of primitive cultures. She is practicing for an unusual night-
club act in which she tells an amusing story, illustrated by a
dazzling series of string patterns that she forms rapidly on
her fingers with a golden cord.

Just as the charm of origami, the Japanese art of paper
folding, lies in the incredible variety of things that can be
done with a single sheet of blank paper, so the charm of
string play lies in the incredible variety of entertaining and
even beautiful things that can be done with one loop of cord.
The string should be about six feet long and knotted at the
ends. The loop is, of course, a model of a simple closed-space

curve. Only the length of the cord and its topological properties remain invariant, so that in a loose sense one can think
of string play as a topological pastime.

There are two basic categories of string play: releases and
catches, and patterns. In stunts of the first category the
string appears to be linked or entangled with an object but,
to everyone's surprise, is suddenly pulled free; or, alternatively, the loop unexpectedly catches on something. For
example, the string is suddenly released from a buttonhole,
or loops are placed around the neck, an arm, a foot—even the
nose—and then mysteriously pulled free. In many releases
the cord is looped once or more around someone's upright
finger and then freed by a series of curious manipulations.
In other releases the string is twisted in a hopeless tangle
around the fingers of the left hand and a tug pulls it free.
There are many variations of an old carnival swindle called
the "garter trick" (it was often performed with a garter in
the days when men wore silk stockings), in which the string
is formed into a pattern on the table; a spectator puts his
finger in one of the loops and then bets on whether the string
will or will not catch on his finger when the swindler pulls
the cord to one side. Of course the operator has subtle ways
of controlling the outcome.

An amusing release that never fails to intrigue all who see
it begins with the string doubled three times to form a small
eight-strand loop about three inches in diameter. Insert your
two forefingers into the loop and rotate it by twirling the
fingers in the manner shown in the drawing numbered *1* in
Figure 90. After twirling for a few moments, stop at the
position indicated by *2*, then touch the tip of each thumb to
the tip of each forefinger as shown in *3*. Lower your right
hand and place the tips of thumbs and forefingers together
as shown in *4*. Note that the right thumb touches the left
finger and the left thumb touches the right finger. (Do not
call attention to this. It is the secret of the trick!) Keeping
thumbs pressed against fingers, raise your right thumb and
left finger as shown in *5*. The loop is now lying on the lower
thumb and finger. At this point a slight forward toss (keeping intact the circle formed by fingers and thumbs) will
throw the loop free of the hands.

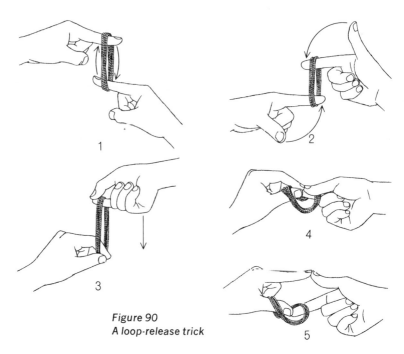

Figure 90
A loop-release trick

Challenge anyone to do what you just did. He will find it astonishingly difficult. Most people assume that thumb touches thumb and finger touches finger. On this assumption it is impossible to free the loop without breaking the circle formed by fingers and thumbs—and such a break is not allowed. Practice until you can do the feat smoothly and rapidly. You will find that you can demonstrate it over and over again without anyone's succeeding in duplicating the moves.

A completely different type of release is that of freeing a ring from the cord. A spectator's upright thumbs hold the string as shown in Figure 91, with the ring riding on both strands. The following is the simplest of many techniques for removing the ring: Place your extended left forefinger over both strands at the point marked *A*. With your right hand pick up the strand nearest you, at point *B*. Draw it upward to the left and place it over the spectator's right thumb (the thumb to your left), moving it from front to back. Curl your

Figure 91
A ring-release trick

left forefinger to retain a firm grip on both strands. Slide the ring to the left as far as you can. Pick up the uppermost strand to the right of the ring, draw it up and to the left and loop it (this time from back to front) over his right thumb.

Pause at this point and ask the spectator to touch the tip of each thumb to the tip of each forefinger. This, you explain, is to make certain that no loop is slipped off either thumb. Grasp the ring with your left hand. Tell him that on the count of three he is to move his hands apart to take up the slack that will form in the cord. When you say "Three," withdraw your left forefinger from the string. As he moves his hands apart the ring comes free. The cord remains on his thumbs exactly as it was at the beginning, without even a twist in it. (As the ring is being released you can slide it along the cord to the right so that it appears to come free near his *left* thumb, where he knows the loop on his thumb is secure.) Children are always delighted by this trick, particularly because it is easily learned and they can show it to friends.

After mastering this release you may wish to try the more sophisticated variation of putting three rings on the cord and removing only the center one. Begin as before, putting the first loop over the spectator's thumb. Slide the first two rings to the left, leaving the third ring near his left thumb. Grasp the upper strand as before, to the right of both rings, but thread it through the first ring before you loop it over his thumb. Hold the middle ring with your right hand and finish as before. Can the reader devise a similar series of manipulations that will put the ring back on the center of the cord again?

Figure 92 shows a ring-and-string release in the form of a puzzle. Fasten a pair of scissors to one end of the cord as shown. The other end is tied to the back of a chair. The problem is to free the scissors without cutting or untying the string. The puzzle is too easy to require an answer at the close of this chapter, although many readers may find it harder than it looks.

A game involving catches and releases, which the reader is unlikely to know because I just invented it, can be played with a loop of cord and a coin. The coin is placed flat on a

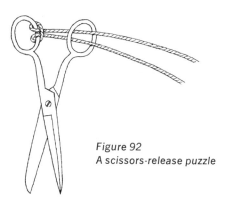

Figure 92
A scissors-release puzzle

table. A player takes the cord by its knot and holds it above the coin so that the loop hangs straight down and touches the coin. He lets it drop in a tangle. Then he places the point of a pencil at any spot on the coin, pushing the point through any opening in the tangle provided he does not alter the position of the cord in any essential way. With one hand keeping the pencil pressed against the coin, he seizes the knot in his other hand and pulls the cord to one side. The probability is high that it will catch on the pencil. He scores 1 if the loop goes around the pencil once, and an additional point for each additional loop. If the cord is wrapped three times around the pencil he scores 3 points. If the cord pulls completely free of the pencil he is docked 5 points. Players take turns and the first to score 30 is the winner.

In the second broad category of string play, various patterns and figures are formed on the hands. This art is part of the folklore of every primitive culture in which string has an important role. For untold generations it has been one of the chief pastimes of the Eskimos, who play it with reindeer sinews and thongs of sealskin. Other cultures in which string figures have reached an advanced stage are those of the North American Indians and of native tribes in Australia, New Zealand, the Caroline Islands, the Hawaiian Islands, the Marshall Islands, the Philippines, New Guinea and the Torres Strait Islands. Over the centuries these natives—particularly the Eskimos—have developed the art to a degree of intricacy that rivals that of paper folding in the Orient and Spain. Thousands of patterns have been invented, some so complex that no one has yet figured out (from the drawings early anthropologists made of completed patterns) the finger manipulations by which they were formed. A native expert can make the patterns with great rapidity. In most cases he uses only his hands, although occasionally he may bring his teeth or toes into play. Often he chants or recites a story while he works.

Most string patterns have acquired names that reflect a fancied resemblance to an animal or some other natural object, and many of these "realistic" figures can be animated in some way. A zigzag flash of lightning appears suddenly between the hands, a sun goes down slowly, a boy climbs a tree, a mouth opens and shuts, two head-hunters battle, a horse gallops, a snake wriggles from hand to hand, a spear is tossed back and forth, a caterpillar is made to crawl along the thigh, a fly vanishes when one tries to squash it between the hands, and so on. Even among the static patterns there are often touches of remarkable realism. A butterfly, for example, has a section of string that coils into a spiral proboscis. In the mystery novel mentioned earlier each murder victim is found with a string pattern on his or her fingers or attached to a piece of cardboard; in each case the pattern symbolizes in some way the character of the victim.

The traditional cat's-cradle game, the only string play widely known among children of Great Britain and the United States, belongs to an interesting class of patterns that demand the cooperation of two players. The string is passed back and forth between the players, forming a new pattern at each transfer. So universal is this pastime that, according to David Riesman (in his book *Individualism Reconsidered*, page 216), "our Army advised soldiers and aviators to always carry a piece of string with them and when downed in a Pacific jungle to start playing cat's cradle if a suspicious native approached; the native would sometimes start to play too."

The literature on string figures is almost as extensive as that on origami. The earliest references are passing mentions of the pastime by a few eighteenth- and nineteenth-century writers. Captain William Bligh, in his log of the voyage of the *Bounty*, 1787–1790 (the period of the famous mutiny), speaks of seeing natives of Tahiti playing with the cord. Charles Lamb recalls string play during his school days. In 1879 the English anthropologist Edward Burnett Tylor called attention to the importance of string figures as culture clues, and in 1888 Franz Boas wrote the first full anthropological description of how a native produces a pattern. A nomenclature and method of describing the making of string figures was published by W. H. R. Rivers and Alfred C. Haddon in 1902. Since then a large number of important papers on string play have appeared in anthropological journals, and many books have been devoted to the subject. There was a time (around 1910) when, if you met a man with a loop of string in his pocket, it was likely he was an anthropologist. Unfortunately, string play turned out to be less significant in cultural anthropological work than it had been thought to be. Today a man with a loop of string is more likely to be an amateur magician.

Most books on string play have long been out of print, but Dover Publications in 1962 reprinted one of the most comprehensive: *String Figures and How to Make Them,* by Caroline Furness Jayne, first published in 1906. This richly illustrated compendium of more than four hundred pages contains detailed instructions for making some one hundred different figures and is an excellent introduction to a fascinating avocation. It is a pity that the art is not more familiar, particularly among teachers of young children, nurses who work with the bedridden and psychiatrists who advise handicraft as therapy.

To whet the reader's appetite I shall explain one of the simplest and most widely known of the diamond patterns. Mrs. Jayne calls it the Osage Diamonds because it was first shown to her by an Osage Indian from Pawhuska, Oklahoma, but it is more commonly known in this country as Jacob's Ladder. The reader is urged to take a six-foot piece of soft cord, knot the ends and see if he can master the figure. With a little practice the diamond pattern can be made in less than ten seconds.

The figure starts, as do most string patterns, with the cord looped over the thumbs and little fingers as shown in the drawing numbered *1* in Figure 93. Put the tip of your right forefinger under the string that crosses your left palm, and with the back of this finger draw the strand to the right. Do the same thing with your left forefinger, putting it between

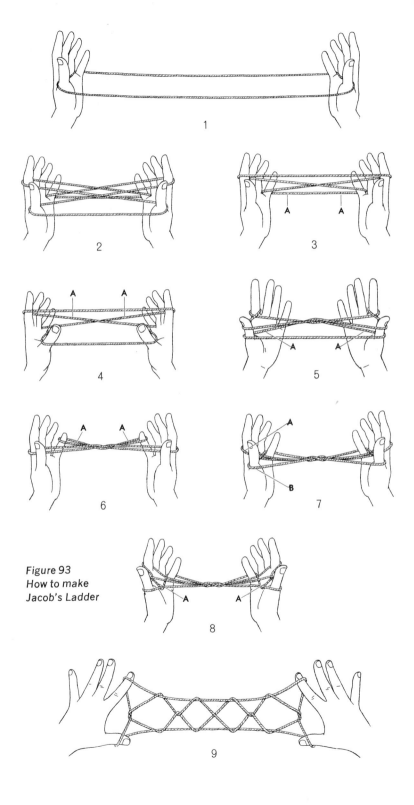

1

2

3

4

5

6

7

Figure 93
How to make
Jacob's Ladder

8

9

the strands now attached to the right forefinger. The cord should appear as it does at *2*. Withdraw your thumbs and pull the string taut (*3*).

Turn your palms away from you to make it easy to put the tips of your thumbs under the farthest strand at the points marked *A* in *3*. With the thumbs, draw this back and under all the other strands to the position shown at *4*. Bend the thumbs over the strand nearest them and with the backs of the thumbs pick up the next strand at the points marked *A* in *4*. Release the little fingers from their loops. The string should appear as it does in *5*.

Bend the little fingers over the strands nearest them and with the backs of the fingers pick up the strands at the points marked *A* in *5*. Release the thumbs. This brings the cord to the position shown at *6*. Bend each thumb over the two strands nearest it and with the backs of the thumbs pick up the next strands at the points marked *A* in *6*. Return the thumbs. The string pattern should now appear as it does at *7*.

With your right thumb and forefinger pick up the string at point *A* (*7*), pull it toward you and place the loop over the left thumb; then take the loop already on the left thumb, holding it at the point marked *B* (*7*), and lift it over the thumb, thereby releasing it. This exchange of loops is known as "Navahoing the loops"—a move that occurs in the making of many string figures. With your left hand Navaho the loops in the same way on the right thumb. (An expert can Navaho both thumbs simultaneously without the help of the other hand, but a beginner had best do it in the manner described.) The string now appears as it does at *8*.

You are ready for the final move. Bend your forefingers, placing their tips down into the small triangles marked *A* in *8*. Withdraw your little fingers from the string, at the same time turning your palms away from you, raising the forefingers as high as you can. (Allow plenty of slack in the string during this final maneuver or the pattern will not open fully.) Draw the cord taut. If the maneuver is done properly, the diamond pattern will form as shown at *9*. This sudden appearance of a pleasing design out of what had appeared to be chaos is one of the delightful features of most string patterns.

Two people who master the figure will find it amusing to produce it cooperatively, the cord then being held by one player's left hand and the other player's right hand. It is not difficult to produce two identical patterns simultaneously in this manner, each on a pair of hands shared by two players. The ultimate test of dexterity is for two players, rapidly and at the same time, to share hands and form two different patterns, but this calls for great skill and coordination.

A puzzle greeting is concealed in Figure 94. This is a poem called "Suicide," written by Louis Aragon, the French writer, during his early association with the surrealist movement. I take it to symbolize life as it appears to the despondent: all its rich variety drained away, leaving only an idiotic ordering of meaningless symbols. In brooding on this poem I have discovered that Aragon unintentionally hid within it a two-word exhortation that, in the light of the nuclear arms race, seems an appropriate message for our time. To decode it place the point of a pencil on a certain letter, then move from letter to adjoining letter, up or down, left or right or diagonally, spelling out the message. (In other words, move like a chess king.) A letter may be counted

a	b	c	d	e	f	
	g	h	i	j	k	l
	m	n	o	p	q	r
	s	t	u	v	w	
		x	y	z		

Figure 94
Louis Aragon's poem "Suicide"

twice to permit such spellings as "stunning" and "no onions." In spite of severe limitations imposed by the sparsity of vowels, it is possible to obtain fairly long phrases: for example, "No point to hide" and "Put UN on top." The two-word phrase I have in mind, however, is remarkably appropriate when addressed to a world about to cut its own throat; moreover, it has a marvelous ambiguity.

ADDENDUM

Jerome Barry, who wrote the mystery novel about string figures, was working for a Manhattan advertising agency when I visited him in 1962. He told me he first became so intrigued by string play that he would carry the loop of cord with him and form the figures in idle moments. He explained to so many people that it all had to do with a mystery novel he was writing that he finally had to write one. About 1950 he used string figures again in a mystery that he wrote for the "Lights Out" television show. The show's leading man, he told me, couldn't master the figures; so they were prepared in advance and coated with glue to give the string a permanently rigid shape. The camera would show the actor making the first string move, shift to a close-up of Barry's hands until the figure was completed, then back to the actor with the glued pattern on his fingers.

A. Richard King, who in 1962 was teaching a fourth-grade class in Carcross, Yukon Territory, Canada, sent me the following letter:

> DEAR MR. GARDNER:
> Osage Diamonds, the word "charge," and a sense of humility are permanent associations in my mind. The whole thing began with your article about string play. . . .
> I am teacher of a fourth-grade class at an Indian Residential School here. This string play seemed a natural device for capturing interest among the children. I had never observed them in any sort of string play. One time I had showed some small ones the Cat's Cradle, elicited some mildly pleased responses, but had

seen no follow-up activity. (These children come from various parts of interior Yukon; have no tribal identification; do not speak a language other than English; are descended principally from Athapaskan-speaking bands of nomadic people who used to be known as Kutchin, Han, or Kaska.)

My own efforts to produce the Osage Diamonds were strenuous and frustrating. After having produced what must have been every possible incorrect variation, I finally arrived at the correct series of manipulations but was quite awkward with the final flip. I discarded the notion of trying to teach it to my children for it was obviously too complicated for them to master.

A month or so later I was droning through a lesson one warm afternoon. We were hung up on the word, "charge," from the spelling lesson. We had done all right with the concept of "attack" and "being responsible for"; and even "credit" was not difficult. But the trouble was distinguishing between "credit" and being charged for something and expected to pay for it immediately.

One of the better girl students, who is more often than not in control of whatever concept we are worrying, was lounging on the base of her spine idly playing with a piece of yarn. A flip of her hand and there was the Osage Diamond! That instant is indelible in my memory. I don't know what I said, but I can still feel my mouth hanging slack. Gently, so as not to make her feel I was going to be punitive, I probed to see where her skill came from.

My surprise was nothing compared to that of the children when I showed interest in such foolishness. Why *everyone* in the class knew *that*! Of course, I lost the group for further school work that day. If the teacher was crazy enough to allow—and appear to like—string tricks, there were plenty to show him. So I saw "Broom" and "Teacup" and "Baby on a Swing" and all the variations of the diamonds over and over for the rest of the day. I doubt that we still have the several concepts of "charge" under control.

They learned the string patterns from other children slightly older. Adults can remember doing such things when they were young but those to whom I have spoken can no longer remember the actual techniques.

They say that they could easily do it again "with a little practice." No particular significance is attached to the string play. It is "just something children always do."

The Osage Diamonds as pictured and described in your article are one of a series these children call simply, "Twos," "Threes," "Fours," etc. They can go up to "Sixes," the number indicating the number of diamonds produced in the completed figure.

Enclosed is a picture of one of our children doing "Twos" and "Fours" and "Broom." You were quite correct in stating that these can easily be done in 10 seconds or less. The variation you offered of two people doing the diamonds together, using one hand apiece, was something new to the children. They quickly mastered this technique and have enjoyed it. . . .

Thank you for a most interesting experience.

ANSWERS

The two-word phrase I had in mind, concealed in Aragon's poem, is "Chin up." It can, of course, be taken in two different ways.

Five readers (J. R. Bruman, Richard Jenney, Alex Schapira, Jane Sichak, Robert Smyth) suggested "Stop, idiots!" (or "Idiots, stop!"). "Stupid idiots" was proposed by Jack Westfall, Herman Arthur, and Richard Jenney. Others included: "Hide idiots" (Ron Edwards), "Join up!" (Marvin Aronson), "Hoping not" (David Harper), "Feint not" (Richard De Long), and "No hoping" (Harmon Goldstone).

Judith M. Hobart spelled out the following telegram from U Thant to President Kennedy and Premier Khrushchev: HINT TO J. F. K., K.: JOIN TO PUT UN ON TOP. IN HOPING, NO POINT; TO HIDING, NO OUT. STUPID IDIOTS, STOP!

Linus Pauling suggested a two-word phrase in which ambiguity is provided by a pun. "My wife and I thought that 'No hiding' would be the solution," he wrote. "Not only is there no way to hide from nuclear war; it is no longer possible for one great nation to give another a hiding."

Two readers, both in Toronto, extracted poems from Aragon's poem. Dennis Burton sent:

Zut!
Chin up John,
You pout?
Gab!
Yup!
Hop not on pont,
JOHN HIP?
No, idiot, no dice.

Nuts!
fed up. fed up Id,
Ide to hide,
chide bag,
ion, pion, pin.

To die?
no point,
top too hot.

The Varsity, the student newspaper at the University of Toronto, asked its readers on February 8, 1963, to see what they could spell in Aragon's poem. On February 15 it printed the following poem by Eleonor Anderson, of the university's Banting Institute:

To the Leaders Who

"put out no opinion."
Snide idiots!
Snide feints, not stopping,
hiding,
chopping.

No point to hide in,
I chide: "Idiots, stop!
Join in stopping, not
to join in hiding."

Hoping not to, I die.

Note that the entire poem can be spelled with an unbroken series of king moves.

CHAPTER EIGHTEEN

O

Curves of
Constant Width

IF AN ENORMOUSLY heavy object has to be moved from one spot to another, it may not be practical to move it on wheels. Axles might buckle or snap under the load. Instead the object is placed on a flat platform that in turn rests on cylindrical rollers. As the platform is pushed forward, the rollers left behind are picked up and put down again in front.

An object moved in this manner over a flat, horizontal surface obviously does not bob up and down as it rolls along. The reason is simply that the cylindrical rollers have a circular cross section, and a circle is a closed curve possessing what mathematicians call "constant width." If a closed convex curve is placed between two parallel lines and the lines are moved together until they touch the curve, the distance between the parallel lines is the curve's "width" in one direction. An ellipse clearly does not have the same width in all directions. A platform riding on elliptical rollers would wobble up and down as it rolled over them. Because a circle has the same width in all directions, it can be rotated between two parallel lines without altering the distance between the lines.

Is the circle the only closed curve of constant width? Most people would say yes, thus providing a sterling example of how far one's mathematical intuition can go astray. Actually there is an infinity of such curves. Any one of them can be the cross section of a roller that will roll a platform as smoothly as a circular cylinder! The failure to recognize such curves can have and has had disastrous consequences in industry. To give one example, it might be thought that the cylindrical hull of a half-built submarine could be tested for circularity by just measuring maximum widths in all directions. As will soon be made clear, such a hull can be monstrously lopsided and still pass such a test. It is precisely for this reason that the circularity of a submarine hull is always tested by applying curved templates.

The simplest noncircular curve of constant width has been named the Reuleaux triangle after Franz Reuleaux (1829–1905), an engineer and mathematician who taught at the Royal Technical High School in Berlin. The curve itself was known to earlier mathematicians, but Reuleaux was the first to demonstrate its constant-width properties. It is easy to construct. First draw an equilateral triangle, *ABC (see Figure 95)*. With the point of a compass at *A*, draw an arc, *BC*. In a similar manner draw the other two arcs. It is obvious that the "curved triangle" (as Reuleaux called it) must have a constant width equal to the side of the interior triangle.

Figure 95

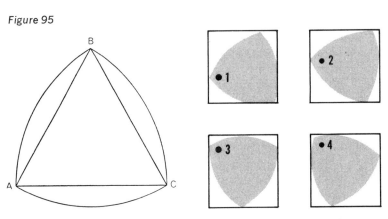

Construction of Reuleaux triangle   Reuleaux triangle rotating in square

If a curve of constant width is bounded by two pairs of parallel lines at right angles to each other, the bounding lines necessarily form a square. Like the circle or any other curve of constant width, the Reuleaux triangle will rotate snugly within a square, maintaining contact at all times with all four sides of the square (*see Figure 95*). If the reader cuts a Reuleaux triangle out of cardboard and rotates it inside a square hole of the proper dimensions cut in another piece of cardboard, he will see that this is indeed the case.

As the Reuleaux triangle turns within a square, each corner traces a path that is almost a square; the only deviation is at the corners, where there is a slight rounding. The Reuleaux triangle has many mechanical uses, but none is so bizarre as the use that derives from this property. In 1914 Harry James Watts, an English engineer then living in Turtle Creek, Pennsylvania, invented a rotary drill based on the Reuleaux triangle and capable of drilling square holes! Since 1916 these curious drills have been manufactured by the Watts Brothers Tool Works in Wilmerding, Pennsylvania. "We have all heard about left-handed monkey wrenches, fur-lined bathtubs, cast-iron bananas," reads one of their descriptive leaflets. "We have all classed these things with the ridiculous and refused to believe that anything like that could ever happen, and right then along comes a tool that drills square holes."

The Watts square-hole drill is shown in Figure 96. At right is a cross section of the drill as it rotates inside the hole it is boring. A metal guide plate with a square opening is

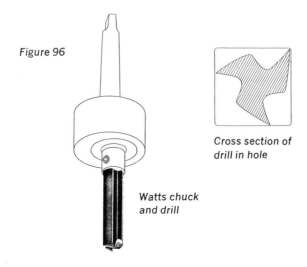

Figure 96

Cross section of
drill in hole

Watts chuck
and drill

first placed over the material to be drilled. As the drill spins within the guide plate, the corners of the drill cut the square hole through the material. As you can see, the drill is simply a Reuleaux triangle made concave in three spots to provide for cutting edges and outlets for shavings. Because the center of the drill wobbles as the drill turns, it is necessary to allow for this eccentric motion in the chuck that holds the drill. A patented "full floating chuck," as the company calls it, does the trick. (Readers who would like more information on the drill and the chuck can check United States patents 1,241,175; 1,241,176; and 1,241,177; all dated September 25, 1917.)

The Reuleaux triangle is the curve of constant width that has the smallest area for a given width (the area is $\frac{1}{2}$ $(\pi - \sqrt{3})\ w^2$, where w is the width). The corners are angles of 120 degrees, the sharpest possible on such a curve. These corners can be rounded off by extending each side of an equilateral triangle a uniform distance at each end (*see Figure 97*). With the point of a compass at A draw arc $DI;$ then widen the compass and draw arc FG. Do the same at the other corners. The resulting curve has a width, in all directions, that is the sum of the same two radii. This of course makes it a curve of constant width. Other symmetrical curves

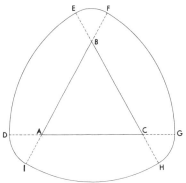

Figure 97
Symmetrical rounded-corner
curve of constant width

of constant width result if you start with a regular pentagon (or any regular polygon with an odd number of sides) and follow similar procedures.

There are ways to draw unsymmetrical curves of constant width. One method is to start with an irregular star polygon (it will necessarily have an odd number of points) such as the seven-point star shown in black in Figure 98. All of these line segments must be the same length. Place the compass point at each corner of the star and connect the two opposite corners with an arc. Because these arcs all have the same radius, the resulting curve (shown in gray) will have constant width. Its corners can be rounded off by the method used before. Extend the sides of the star a uniform distance at all points (shown with broken lines) and then join the ends of the extended sides by arcs drawn with the compass point at each corner of the star. The rounded-corner curve, which is shown in black, will be another curve of constant width.

Figure 99 demonstrates another method. Draw as many straight lines as you please, all mutually intersecting. Each arc is drawn with the compass point at the intersection of the two lines that bound the arc. Start with any arc, then proceed around the curve, connecting each arc to the preceding one. If you do it carefully, the curve will close and will have constant width. (Proving that the curve must close and have constant width is an interesting and not difficult exercise.) The preceding curves were made up of arcs of no more than two different circles, but curves drawn in this way may have arcs of as many different circles as you wish.

A curve of constant width need not consist of circular arcs. In fact, you can draw a highly arbitrary convex curve from the top to the bottom of a square and touching its left side (*arc* ABC *in Figure 99*), and this curve will be the left

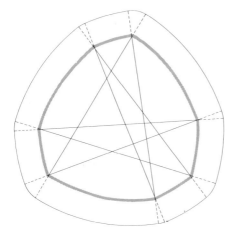

Figure 98
Star-polygon method of
drawing a curve of constant
width

Figure 99

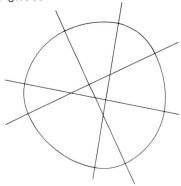

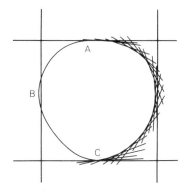

Crossed-lines method   Random curve and tangents

side of a uniquely determined curve of constant width. To
find the missing part, rule a large number of lines, each
parallel to a tangent of arc ABC and separated from the
tangent by a distance equal to the side of the square. This
can be done quickly by using both sides of a ruler. The original square must have a side equal to the ruler's width. Place
one edge of the ruler so that it is tangent to arc ABC at one
of its points, then use the ruler's opposite edge to draw a
parallel line. Do this at many points, from one end of arc
ABC to the other. The missing part of the curve is the envelope of these lines. In this way you can obtain rough outlines of an endless variety of lopsided curves of constant
width.

It should be mentioned that the arc ABC cannot be completely arbitrary. Roughly speaking, its curvature must not
at any point be less than the curvature of a circle with a
radius equal to the side of the square. It cannot, for example, include straight-line segments. For a more precise statement on this, as well as detailed proofs of many elementary
theorems involving curves of constant width, the reader is
referred to the excellent chapter on such curves in *The Enjoyment of Mathematics,* by Hans Rademacher and Otto
Toeplitz.

If you have the tools and skills for woodworking, you
might enjoy making a number of wooden rollers with cross
sections that are various curves of the same constant width.
Most people are nonplused by the sight of a large book rolling horizontally across such lopsided rollers without bobbing
up and down. A simpler way to demonstrate such curves is

to cut from cardboard two curves of constant width and nail them to opposite ends of a wooden rod about six inches long. The curves need not be of the same shape, and it does not matter exactly where you put each nail as long as it is fairly close to what you guess to be the curve's "center." Hold a large, light-weight empty box by its ends, rest it horizontally on the attached curves and roll the box back and forth. The rod wobbles up and down at both ends, but the box rides as smoothly as it would on circular rollers!

The properties of curves of constant width have been extensively investigated. One startling property, not easy to prove, is that the perimeters of all curves with constant width n have the same length. Since a circle is such a curve, the perimeter of any curve of constant width n must of course be πn, the same as the circumference of a circle with diameter n.

The three-dimensional analogue of a curve of constant width is the solid of constant width. A sphere is not the only such solid that will rotate within a cube, at all times touching all six sides of the cube; this property is shared by all solids of constant width. The simplest example of a nonspherical solid of this type is generated by rotating the Reuleaux triangle around one of its axes of symmetry (*see Figure 100 left*). There is an infinite number of others. The

Figure 100
Two solids of
constant width

solids of constant width that have the smallest volumes are derived from the regular tetrahedron in somewhat the same way the Reuleaux triangle is derived from the equilateral triangle. Spherical caps are first placed on each face of the tetrahedron, then it is necessary to alter three of the edges slightly. These altered edges may either form a triangle or radiate from one corner. The solid at the right of Figure 100 is an example of a curved tetrahedron of constant width.

Since all curves of the same constant width have the same

perimeter, it might be supposed that all solids of the same constant width have the same surface area. This is not the case. It was proved, however, by Hermann Minkowski (the Polish mathematician who made such great contributions to relativity theory) that all *shadows* of solids of constant width (when the projecting rays are parallel and the shadow falls on a plane perpendicular to the rays) are curves of the same constant width. All such shadows have equal perimenters (π times the width).

Michael Goldberg, an engineer with the Bureau of Naval Weapons in Washington, has written many papers on curves and solids of constant width, and is recognized as being this country's leading expert on the subject. He has introduced the term "rotor" for any convex figure that can be rotated inside a polygon or polyhedron while at all times touching every side or face.

The Reuleaux triangle is, as we have seen, the rotor of least area in a square. The least-area rotor for the equilateral triangle is shown at the left of Figure 101. This lens-shaped figure (it is not, of course, a curve of constant width) is formed with two 60-degree arcs of a circle having a radius equal to the triangle's altitude. Note that as it rotates its corners trace the entire boundary of the triangle, with no rounding of corners. Mechanical reasons make it difficult to rotate a drill based on this figure, but Watts Brothers makes other drills, based on rotors for higher-order regular polygons, that drill sharp-cornered holes in the shape of pentagons, hexagons and even octagons. In three-space, Goldberg has shown, there are nonspherical rotors for the regular tetrahedron and octahedron, as well as the cube, but none for the regular dodecahedron and icosahedron. Almost no work has been done on rotors in dimensions higher than three.

Closely related to the theory of rotors is a famous problem named the Kakeya needle problem after the Japanese mathe-

Figure 101

Least-area rotor in equilateral triangle

Line rotated in deltoid curve

matician Sôichi Kakeya, who first posed it in 1917. The problem is as follows: What is the plane figure of least area in which a line segment of length 1 can be rotated 360 degrees? The rotation obviously can be made inside a circle of unit diameter, but that is far from the smallest area.

For many years mathematicians believed the answer was the deltoid curve shown at the right of Figure 101, which has an area exactly half that of a unit circle. (The deltoid is the curve traced by a point on the circumference of a circle as it rolls around the inside of a larger circle, when the diameter of the small circle is either one third or two thirds that of the larger one.) If you break a toothpick to the size of the line segment shown, you will find by experiment that it can be rotated inside the deltoid as a kind of one-dimensional rotor. Note how its end points remain at all times on the deltoid's perimeter.

In 1927, ten years after Kakeya popped his question, the Russian mathematician Abram Samoilovitch Besicovitch (then living in Copenhagen) dropped a bombshell. He proved that the problem had no answer. More accurately, he showed that the answer to Kakeya's question is that there is *no* minimum area. The area can be made as small as one wants. Imagine a line segment that stretches from the earth to the moon. We can rotate it 360 degrees within an area as small as the area of a postage stamp. If that is too large, we can reduce it to the area of Lincoln's nose on a postage stamp.

Besicovitch's proof is too complicated to give here (see references in bibliography), and besides, his domain of rotation is not what topologists call simply connected. For readers who would like to work on a much easier problem: What is the smallest *convex* area in which a line segment of length 1 can be rotated 360 degrees? (A convex figure is one in which a straight line, joining any two of its points, lies entirely on the figure. Squares and circles are convex; Greek crosses and crescent moons are not.)

ADDENDUM

Although Watts was the first to acquire patents on the process of drilling square holes with Reuleaux-triangle drills, the procedure was apparently known earlier. Derek Beck, in London, wrote that he had met a man who recalled having used such a drill for boring square holes when he was an apprentice machinist in 1902, and that the practice then seemed to be standard. I have not, however, been able to learn anything about the history of the technique prior to Watts's 1917 patents.

ANSWERS

What is the smallest convex area in which a line segment of length 1 can be rotated 360 degrees? The answer: An equilateral triangle with an altitude of 1. (The area is one third the square root of 3.)

Any figure in which the line segment can be rotated obviously must have a width at least equal to 1. Of all convex figures with a width of 1, the equilateral triangle of altitude 1 has the smallest area. (For a proof of this the reader is referred to *Convex Figures,* by I. M. Yaglom and V. G. Boltyanskii, pages 221–22.) It is easy to see that a line segment of length 1 can in fact be rotated in such a triangle (*see Figure 102*).

The deltoid curve was believed to be the smallest simply-connected area solving the problem until 1963 when a smaller area was discovered independently by Melvin Bloom and I. J. Schoenberg. (See H.S.M. Coxeter, *Twelve Geometric Essays,* [Carbondale and Edwardsville: Southern Illinois University Press, 1968], page 231.)

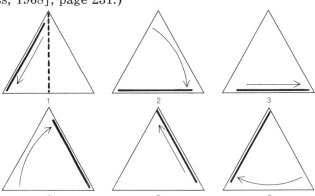

Figure 102   *The answer to needle-turning problem*

CHAPTER NINETEEN

O

Rep-Tiles:
Replicating Figures
on the Plane

ONLY THREE REGULAR POLYGONS—the equilateral triangle, the square and the regular hexagon—can be used for tiling a floor in such a way that identical shapes are endlessly repeated to cover the plane. But there is an infinite number of irregular polygons that can provide this kind of tiling. For example, a triangle of any shape whatever will do the trick. So will any four-sided figure. The reader can try the following test. Draw an irregular quadrilateral (it need not even be convex, which is to say that it need not have interior angles that are all less than 180 degrees) and cut twenty or so copies from cardboard. It is a pleasant task to fit them all together snugly, like a jigsaw puzzle, to cover a plane.

There is an unusual and less familiar way to tile a plane. Note that each trapezoid at the top of Figure 103 has been divided into four smaller trapezoids that are exact replicas of the original. The four replicas can, of course, be divided in the same way into four still smaller replicas, and this can be continued to infinity. To use such a figure for tiling we have only to proceed to infinity in the opposite direction: we put together four figures to form a larger model, four of which will in turn fit together to make a still larger one. The British mathematician Augustus De Morgan summed up this sort of situation admirably in the following jingle, the first four lines of which paraphrase an earlier jingle by Jonathan Swift:

> *Great fleas have little fleas*
> *Upon their backs to bite 'em,*
> *And little fleas have lesser fleas,*
> *And so ad infinitum.*
> *The great fleas themselves, in turn,*
> *Have greater fleas to go on;*
> *While these again have greater still,*
> *And greater still, and so on.*

Until recently not much was known about polygons that have this curious property of making larger and smaller copies of themselves. In 1962 Solomon W. Golomb, who was then on the staff of the Jet Propulsion Laboratory of the California Institute of Technology and is now professor of electrical engineering at the University of Southern California, turned his attention to these "replicating figures"—or "rep-tiles," as he calls them. The result was three privately issued papers that lay the groundwork for a general theory of polygon "replication." These papers, from which almost all that follows is extracted, contain a wealth of material of great interest to the recreational mathematician.

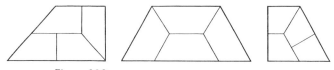

Figure 103
Three trapezoids that have a replicating order of 4

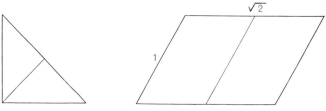

The only known rep-2 polygons

In Golomb's terminology a replicating polygon of order k is one that can be divided into k replicas congruent to one another and similar to the original. Each of the three trapezoids in Figure 103, for example, has a replicating order of 4, abbreviated as rep-4. Polygons of rep-k exist for any k, but they seem to be scarcest when k is a prime and to be most abundant when k is a square number.

Only two rep-2 polygons are known: the isosceles right triangle and the parallelogram with sides in the ratio of 1 to the square root of 2 (*see bottom of Figure 103*). Golomb found simple proofs that these are the only possible rep-2 triangles and quadrilaterals, and there are no other convex rep-2 polygons. The existence of concave rep-2 polygons appears unlikely, but so far their nonexistence has not been proved.

The interior angles of the parallelogram can vary without affecting its rep-2 property. In its rectangular form the rep-2 parallelogram is almost as famous in the history of art as the "golden rectangle," discussed in the *Second Scientific American Book of Mathematical Puzzles and Diversions.* Many medieval and Renaissance artists (Albrecht Dürer, for instance) consciously used it for outlining rectangular pictures. A trick playing card that is sometimes sold by street-corner pitchmen exploits this rectangle to make the ace of diamonds seem to diminish in size three times (*see Figure 104*). Under cover of a hand movement the card is secretly folded in half and turned over to show a card exactly half the size of the preceding one. If each of the three smaller aces is a rectangle similar to the original, it is easy to show that only a 1-by-$\sqrt{2}$ rectangle can be used for the card. The rep-2 rectangle also has less frivolous uses. Printers who

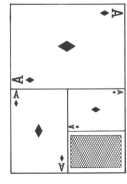

Figure 104
A trick diminishing
card based on
the rep-2 rectangle

wish to standardize the shape of the pages in books of various sizes find that in folio, quarto or octavo form it produces pages that are all similar rectangles.

The rep-2 rectangle belongs to the family of parallelograms shown in the top illustration of Figure 105. The fact that a parallelogram with sides of 1 and $\sqrt{k}$ is always rep-k proves that a rep-k polygon exists for any k. It is the only known example, Golomb asserts, of a family of figures that exhibit all the replicating orders. When k is 7 (or any prime greater than 3 that has the form $4n - 1$), a parallelogram of this family is the only known example. Rep-3 and rep-5 triangles exist. Can the reader construct them?

A great number of rep-4 figures are known. Every triangle is rep-4 and can be divided as shown in the second illustration from the top of Figure 105. Among the quadrilaterals, any parallelogram is rep-4, as shown in the same illustration. The three trapezoids in the top illustration of Figure 103 are the only other examples of rep-4 quadrilaterals so far discovered.

Only one rep-4 pentagon is known: the sphinx-shaped figure in the third illustration from the top of Figure 105. Golomb was the first to discover its rep-4 property. Only the outline of the sphinx is given so that the reader can have the pleasure of seeing how quickly he can dissect it into four smaller sphinxes. (The name "sphinx" was given to this figure by T. H. O'Beirne, of Glasgow.)

There are three known varieties of rep-4 hexagons. If any rectangle is divided into four quadrants and one quadrant is thrown away, the remaining figure is a rep-4 hexagon. The hexagon at the right at the bottom of Figure 105 shows the dissection (familiar to puzzlists) when the rectangle is a square. The other two examples of rep-4 hexagons (each of which can be dissected in more than one way) are shown at the middle and left in the same illustration.

No other example of a standard polygon with a rep-4 property is known. There are, however, "stellated" rep-4 polygons (a stellated polygon consists of two or more polygons joined at single points), two examples of which, provided by Golomb, are shown at the top of Figure 106. In the first example a pair of identical rectangles can be substituted for the squares. In addition, Golomb has found three

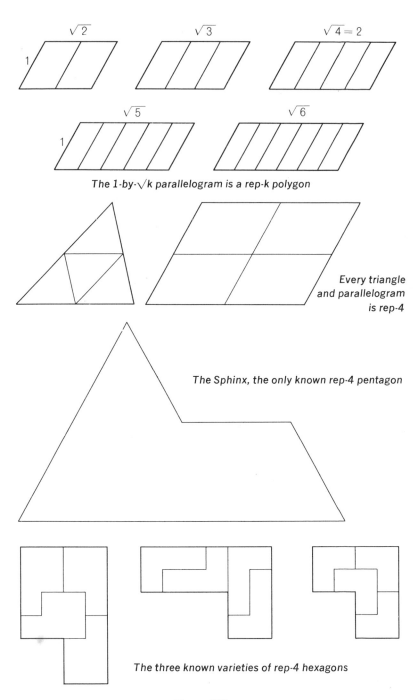

The 1-by-√k parallelogram is a rep-k polygon

Every triangle and parallelogram is rep-4

The Sphinx, the only known rep-4 pentagon

The three known varieties of rep-4 hexagons

Figure 105

nonpolygonal figures that are rep-4, although none is con-
structible in a finite number of steps. Each of these figures,
shown at the left in the bottom illustration of Figure 106, is
formed by adding to an equilateral triangle an endless

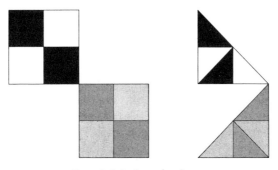

Two stellated rep-4 polygons

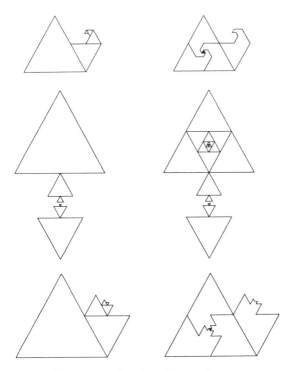

Three examples of rep-4 nonpolygons

Figure 106

series of smaller triangles, each one fourth the size of its predecessor. In each case four of these figures will fit together to make a larger replica, as shown at the right in the same illustration. (There is a gap in each replica because the original cannot be drawn with an infinitely long series of triangles.)

It is a curious fact that every known rep-4 polygon of a standard type is also rep-9. The rep-4 Nevada-shaped trapezoid of Figure 107 can be dissected into nine replicas in many ways, only one of which is shown. (Can the reader dissect each of the other rep-4 polygons, not counting the stellated and infinite forms, into nine replicas?) The converse is also true: All known standard rep-9 polygons are also rep-4.

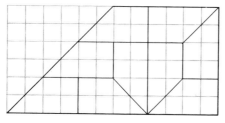

Figure 107
Every rep-4 polygon is also rep-9

Three interesting examples of stellated rep-9 polygons, discovered and named by Golomb, are shown in Figure 108. None of these polygons is rep-4.

Any method of dividing a 4 × 4 checkerboard along grid lines into four congruent parts (as discussed in Chapter 16) provides a figure that is rep-16. It is only necessary to put four of the squares together to make a replica of one of the parts as in Figure 109. In a similar fashion, a 6 × 6 checkerboard can be quartered in many ways to provide rep-36 figures, and an equilateral triangle can be divided along triangular grid lines into rep-36 polygons (*see Figure 110*). All of these examples illustrate a simple theorem, which Golomb explains as follows:

Consider a figure P that can be divided into two or more congruent figures, not necessarily replicas of P. Call the smaller figure Q. The number of such figures is the "multiplicity" with which Q divides P. For example, in Figure 110 the three hexagons divide the triangle with a multiplicity of

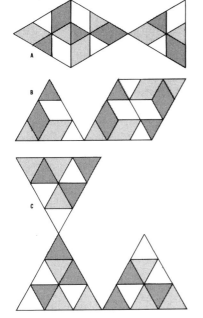

Figure 108
Stellated rep-9 polygons: The Fish (a),
The Bird (b) and The Ampersand (c)

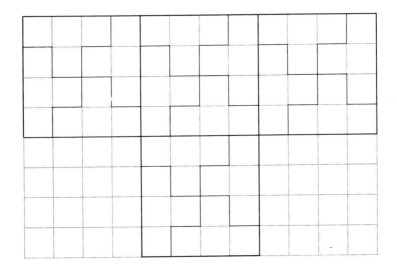

Figure 109
A rep-16 octagon

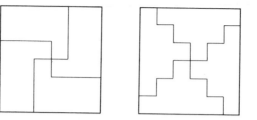

Figure 110
Three rep-36 polygons

3 and small equilateral triangles will divide each hexagon with a multiplicity of 12. The product of these two multiplicities (3 × 12) gives a replicating order for both the hexagon and the equilateral triangle: 36 of the hexagonal figures will form a larger figure of similar shape, and 36 equilateral triangles will form a larger equilateral triangle. In more formal language: If P and Q are two shapes such that P divides Q with a multiplicity of s, and Q divides P with a multiplicity of t, then P and Q are both replicating figures of order st ($s \times t$). Of course, each figure can have lower replicating orders as well. In the example given, the equilateral triangle, in addition to being rep-36, is also rep-4, rep-9, rep-16 and rep-25.

When P and Q are similar figures, it follows from the above theorem that if the figure has a replicating order of k, it will also be rep-k^2, rep-k^3, rep-k^4 and so on for all powers of k. Similarly, if a figure is both rep-s and rep-t, it will also be rep-st.

The principle underlying all of these theorems can be extended as follows. If P divides Q with a multiplicity of s, and Q divides R with a multiplicity of t, and R divides P with a multiplicity of u, then P and Q and R are each rep-stu. For instance, each of the hexominoes in Figure 111 will divide

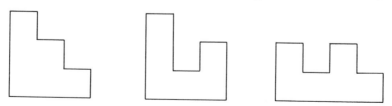

Figure 111
Three rep-144 polygons

a 3 × 4 rectangle with a multiplicity of 2. The 3 × 4 rectangle in turn divides a square with a multiplicity of 12, and the square divides any one of the three original shapes with a multiplicity of 6. Consequently the replicating order of each hexomino is 2 × 12 × 6, or 144. It is conjectured that none of the three has a lower replicating order.

Golomb has noted that every known polygon of rep-4, including the stellated polygons, will divide a parallelogram with a multiplicity of 2. In other words, if any known rep-4 polygon is replicated, the pair can be fitted together to form a parallelogram! It is conjectured, but not yet proved, that this is true of all rep-4 polygons.

An obvious extension of Golomb's pioneer work on replication theory (of which only the most elementary aspects have been detailed here) is into three or even higher dimensions. A trivial example of a replicating solid figure is the cube: it obviously is rep-8, rep-27 and so on for any order that is a cubical number. Other trivial examples result from giving plane replicating figures a finite thickness, then forming layers of larger replicas to make a model of the original solid. Less trivial examples certainly exist; a study of them might lead to significant results.

In addition to the problems already posed, here are two unusual dissection puzzles closely related to what we have been considering (*see Figure 112*). First the easier one: Can

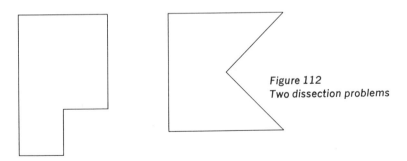

Figure 112
Two dissection problems

the reader divide the hexagon (*left*) into two congruent stellated polygons? More difficult: Divide the pentagon (*right*) into four congruent stellated polygons. In neither case are the polygons similar to the original figure.

ADDENDUM

The conjecture that the three polygons shown in Figure 111 could not be cut into fewer than 144 replicas turned out to be true only of the two end ones. Mark A. Mandel, New York City, then fourteen years old, wrote to show how the middle polygon could be cut into 36 replicas. Readers may enjoy searching for the pattern.

Ralph H. Hinrichs, Phoenixville, Pennsylvania, discovered that if the middle hexagon at the bottom of Figure 105 is dissected in a slightly different way (the pattern within each rectangle is mirror-reflected), the entire figure can undergo an infinite number of affine transformations (the 90-degree exterior angle taking any acute or obtuse value) to provide an infinity of rep-4 hexagons. Only when the angle is 90 degrees is the figure also rep-9, thus disproving an early guess that all rep-4 standard polygons are rep-9 and vice versa.

More recent results in the field are given in the last three references cited in the bibliography for this chapter.

ANSWERS

The problem of dissecting the sphinx is shown in Figure 113, top. The next two illustrations show how to construct rep-3 and rep-5 triangles. The bottom illustration gives the solution to the two dissection problems involving stellated polygons. The first of these can be varied in an infinite number of ways; the solution shown here is one of the simplest.

The second solution is an old-timer. Sam Loyd, in his puzzle column in *Woman's Home Companion* (October 1905) points out that the figure is similar to the one shown here in the lower right corner of Figure 105 in that one fourth of a square is missing from each figure. He writes that he spent a year trying to cut the mitre shape into four congruent parts, each simply connected, but was unable to do better than the solution reproduced here. It can be found in many old puzzle books antedating Loyd's time.

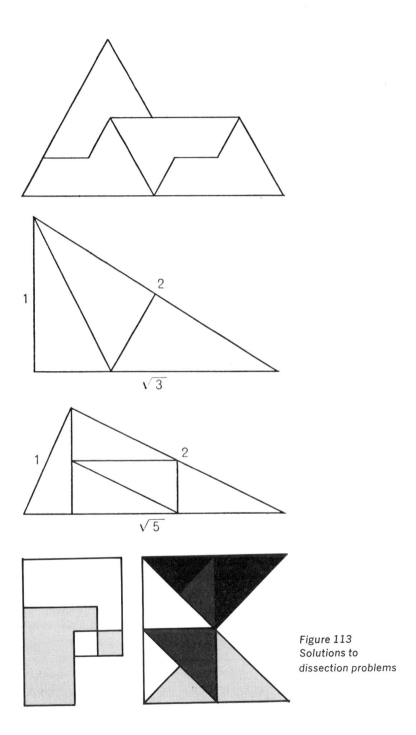

1

2

$\sqrt{3}$

1

2

$\sqrt{5}$

Figure 113
Solutions to
dissection problems

CHAPTER TWENTY

O

Thirty-Seven
Catch Questions

HERE IS A COLLECTION of thirty-seven short problems, presented in the hope of "catching" as many readers as possible. Every problem conceals some sort of joke. Only a few are mathematically significant. The reader is urged, however, not to peek at the solutions until he has made at least a semiserious attempt to answer as many of the questions as possible.

1. Three Navaho women sit side by side on the ground. The first woman, who is sitting on a goatskin, has a son who weighs 140 pounds. The second woman, who is sitting on a deerskin, has a son who weighs 160 pounds. The third woman, who weighs 300 pounds, is sitting on a hippopotamus skin. What famous geometric theorem does this symbolize?

2. A tired physicist went to bed at ten o'clock one night after setting his alarm clock for noon the following day. When the alarm woke him, how many hours had he slept?

3. Joe throws an ordinary die, then Moe throws the same die. What is the probability that Joe will throw a higher number than Moe?

Figure 114
A die is thrown by Moe and Joe

4. What is the exact opposite of "not in"?

5. On level ground a 10-foot pole stands a certain distance from a 15-foot pole (*see Figure 115*). If lines are drawn from the top of each pole to the bottom of the other as shown, the lines intersect at a point six feet above the ground. What is the distance between the poles?

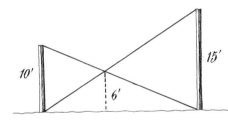

Figure 115
What is the distance between the poles?

6. "How much will one cost?"
"Twenty cents," replied the clerk in the hardware store.
"And how much will twelve cost?"
"Forty cents."
"Okay. I'll take nine hundred and twelve."
"That will be sixty cents."
What was the customer buying?

7. A triangle has sides of 13, 18 and 31 inches. What is the triangle's area?

8. What familiar English word is invariably pronounced wrong by every mathematician at the Institute for Advanced Study in Princeton, New Jersey?

9. John Kennedy was born in 1917. He became president in 1960. His age in 1963 was 46 and he had been in office 3 years. The sum of these four numbers is 3,926. Charles de Gaulle was born in 1890. He became president of France in 1958. His age in 1963 was 73 and he had been in office 5

years. The sum of these four numbers also is 3,926. Can you explain this remarkable coincidence?

10. What angle is made by the two dotted lines on the cube in Figure 116?

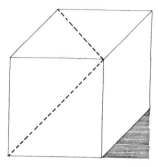

Figure 116
What angle is made by the dotted lines?

11. Rearrange the letters of NEW DOOR to make one word.

12. The edge of a reservoir is a perfect circle. A fish starts at a point on the edge and swims due north for 600 feet, which takes him to the edge again. He then swims due east, reaching the edge after going 800 feet. What is the reservoir's diameter?

13. A statistician gave mathematical tests to everyone who lived in a village of 6,000 people and at the same time measured the lengths of their feet. He found a strong correlation between mathematical ability and foot size. Explain.

14. Roy G. Biv, of Rainbow, Oregon, wants to know what familiar continuum is expressed by the following words: flushed, New Jersey town, cowardly, naïve, depressed, dyestuff, shrinker.

Figure 117
How to measure diameter as the fish swims

15. Write a simple formula with only the one variable, x, such that when any positive integer is substituted for x, the formula is sure to give a prime number.

16. A man wishes to build a house on a large triangular plot of ground, then to construct three straight roads, each leading from the house to a side of the triangle and each road perpendicular to the side. The triangle is equilateral. Where should he place his house in order to minimize the sum of the lengths of the three roads?

Figure 118
A house builder's triangular
problem

17. Divide 50 by 1/2 and add 3. What is the result?

18. In the following line of letters cross out six letters so that the remaining letters, without altering their sequence, will spell a familiar English word:

B S A I N X L E A T N T E A R S

19. A topologist bought seven doughnuts and ate all but three. How many were left?

20. In going over his books one day a bookkeeper for a toy company noticed that the word "balloon" had two sets of double letters, one following the other. "I wonder," he said to himself, "if there is an English word containing *three* sets of double letters, one right after the other." Such a word appears on this page. Can you find it?

Figure 119
The bookkeeper and his balloon

21. The dotted lines in Figure 120 are bisectors of the two base angles of a triangle. They intersect at right angles. Leo Moser of the University of Alberta asks: If the base of the triangle is 10 inches, what is its altitude?

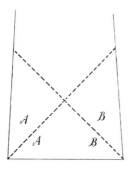

Figure 120
What is the altitude of the triangle?

22. How many months have 30 days?

23. Mrs. Smith wants to stop smoking after she finishes her last remaining nine cigarettes. She can make a new cigarette by wrapping three butts in a piece of cigarette paper. If she uses this technique as many times as she can, how many cigarettes can she smoke before she finally quits?

Figure 121
Mrs. Smith smokes her last cigarettes

24. The following limerick was composed by Leigh Mercer of London. Can you read it correctly?

1,264,853,971.2758463

25. "Here are three pills," a doctor says to you. "Take one every half-hour." You comply. How long will your pills last?

Figure 122
A tennis player moves to the next round

26. One hundred and thirty-seven men have signed up for an elimination tennis tournament. All players are to be paired for the first round, but because 137 is an odd number one player gets a bye, which promotes him to the next round. The pairing continues on each round, with a bye to any player left over. If the schedule is planned so that a minimum number of matches is required to determine the champion, how many matches must be played?

27. Find a word of ten letters that can be typed by using only the top row of letters on a typewriter.

28. A box contains two United States coins that together total 55 cents. One is not a nickel. What are the coins?

29. A fish weighs 20 pounds plus half its own weight. How much does it weigh?

Figure 123
A fish is weighed

30. The following telegram was recently composed by Roger Angell, a writer on the staff of *The New Yorker:*

"MARGE, LET DAM DOGS IN. AM ON SATIRE; VOW I AM CAIN, AM ON SPOT, AM A JAP SNIPER. RED, RAW MURDER ON GI!

IGNORE DRUM . . . WARDER REPINS PAJAMA TOPS . . . NO
MANIAC, MA! IWO VERITAS: NO MAN IS GOD.—MAD TELEGRAM."
What is so remarkable about this message?

31. D. G. Prinz, a mathematician with Ferranti Ltd. in
Manchester, England, discovered the following symmetrical
equation:

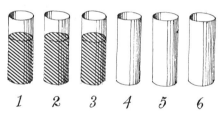

What is the value of x? (Hint: Each set of "III" can be
interpreted in three different ways.)

32. Arrange six glasses in a row as shown in Figure 124.
The first three glasses are filled with water, the last three are
empty. By moving one glass only, change the arrangement
so that the glasses alternate empty with full.

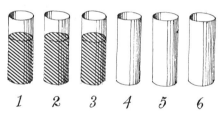

Figure 124
Move one glass to alternate empty and full

33. A wheel has ten spokes. How many spaces does it have
between spokes?

34. "The number of words in this sentence is nine." The
sentence just quoted is obviously true. The opposite of a true
statement is usually false. Give a sentence that says the exact
opposite of the quoted sentence but is nevertheless true.

35. Two girls were born on the same day of the same
month in the same year of the same parents, yet they were
not twins. Explain.

36. If someone says to you, "I'll bet you a dollar that if you give me five dollars I'll give you a hundred dollars in exchange," would that be a good bet to take?

Figure 125
How to lose $4

37. O. Henry's famous short story, "The Gift of the Magi," opens as follows: "One dollar and eighty-seven cents. That was all. And sixty cents of it was in pennies." Is there anything mathematically wrong here?

ANSWERS

1. The squaw on the hippopotamus is equal to the sons of the squaws on the other two hides.

2. Two hours.

3. 5/12. The probability that both will throw the same number is 1/6, therefore the probability that one will throw higher than the other is 5/6 or 10/12. This is halved to give the probability that Joe will get a higher number than Moe.

4. "In."

5. Any distance. The height of the intersection is equal to the product of the heights of the two poles divided by their sum.

6. House numbers.

7. Zero.

8. "Wrong."

9. Any date added to the number of years since that date

will total the current year. Two such totals will be twice the current year.

10. Sixty degrees. Joining the ends of the two lines completes an equilateral triangle.

11. ONE WORD.

William T. Walsh, of the Brookhaven National Laboratory, Upton, Long Island, New York, wrote that before tackling my thirty-seven joke problems he had read, in the same issue of *Scientific American,* an article on the psychology of problem solving. "Hence," he wrote, "I could not read any of the problems you proposed without first examining my attitude to see what particular psychological 'set' I had assumed." When he came to the ONE WORD problem he decided that since nothing was said about the orientation of each letter when they were rearranged, he could turn the W upside down to arrive at the following unique solution: DOORMEN.

12. A thousand feet. The fish makes a right-angle turn. A right angle, with its vertex on the circumference of a circle, intersects the circumference at the end points of a diameter. The diameter is therefore the hypotenuse of a right triangle with sides of 600 and 800 feet.

13. "Everyone" included babies and children.

14. The spectrum of visible light: red, orange, yellow, green, blue, indigo, violet. Roy G. Biv, a mnemonic acronym for the spectrum, is really Stephen Barr of Woodstock, New York.

15. There are many such formulas: $2 + 1^x$, $0^x + 3$, $2 + x/x$, and so on.

16. Anywhere. The sum of the three paths is a constant equal to the triangle's altitude.

17. 103.

18. After crossing out SIX LETTERS, the remaining letters spell BANANA.

A surprising "better" solution, just as legitimate if not more so, was discovered by readers at Conductron Corporation, Ann Arbor, Michigan, and reported to me by Robert E. Machol. The six letters SAINXL are removed wherever they appear, leaving the word BETTER.

19. Three.

20. Bookkeeper. Olin Jerome Ferguson and Leo Moser each called my attention to *subbookkeeper* (listed in *Webster's New International Dictionary*, 2d edition, 1942, page 2507) which has four doublets in a row. "What a boob the bookkeeper will think he is," wrote Peter F. Arvedson, "when he finds out there is an English word with *five* sets of successive double letters that describe him more completely: *boobbookkeeper*." Stephen Barr, when I told him about this, added a sixth doublet with *subboobbookkeeper*.

21. Infinity. Angles *a* and *b* sum to 90 degrees. The two base angles of the triangle (*2a* and *2b*) sum to 180 degrees. Therefore the top angle of the triangle must be 0 degrees, with the sides of the triangle parallel, meeting at infinity.

22. All but February.

23. Thirteen. Pierre Basset, Ekkehard Künzell and Mel Stover were three readers who thought that Mrs. Smith's procedure involved an inexcusable waste of that final butt. It would have been better, each said, had she started with a set of ten cigarettes. After consuming fourteen cigarettes she would be left with two butts. She could find a third butt in an ashtray, smoke her fifteenth and last cigarette, then replace the butt where she found it.

24. One thousand two hundred and sixty-
   Four million eight hundred and fifty-
      Three thousand nine hun-
      Dred and seventy-one
   Point two seven five eight four six three.

25. One hour.

26. Because 136 players must be eliminated, there must be 136 matches.

27. Typewriter. There are many other ten-letter words, and a few even longer ones. See "Typewriter Words" by Dmitri Borgmann in *Language on Vacation* (New York: Scribner's, 1965), pages 171–73.

28. A fifty-cent piece and a nickel. The fifty-cent piece is not a nickel.

29. Forty pounds.

30. The telegram is a palindrome, reading the same backward and forward.

31.

$$X = \frac{111}{3} = 37.$$

In the fraction, the III above the line is in the decimal system, the III below is a Roman numeral. The next III is also Roman and the last III is in the binary system.

Two readers, Frieda Herman and Joel Herskowitz, each proposed a different interpretation. A vertical bar on each side of a real number indicates the absolute value of that number; that is, its value without regard to sign. The equation, therefore, can be taken to mean that x equals the absolute value of 1 divided by the absolute value of 1, which in turn equals the absolute value of 1 multiplied by the absolute value of 1.

32. Pick up the second glass, pour its contents into the fifth glass, replace the second glass.

33. Ten.

34. "The number of words in this sentence, is not nine."

35. They were in a set of triplets.

36. No. He can take your $5, say "I lose," and hand you his $1. You win the bet but lose $4.

37. No. At the time O. Henry wrote this story the United States still had three-cent pieces in circulation. (They were minted as late as 1889.) Two-cent pieces were discontinued in 1873, but remained in circulation for many years after that. One two-cent piece or four three-cent pieces would explain O. Henry's statement.

BIBLIOGRAPHY

O

1. THE PARADOX OF THE UNEXPECTED HANGING

D. J. O'Connor, "Pragmatic Paradoxes," *Mind*, Vol. 57, July 1948, pages 358–59.

L. Jonathan Cohen, "Mr. O'Connor's 'Pragmatic Paradoxes,'" *Mind*, Vol. 59, January 1950, pages 85–87.

Peter Alexander, "Pragmatic Paradoxes," *Mind*, Vol. 59, October 1950, pages 536–38.

Michael Scriven, "Paradoxical Announcements," *Mind*, Vol. 60, July 1951, pages 403–7.

D. J. O'Connor, "Pragmatic Paradoxes and Fugitive Propositions," *Mind*, Vol. 60, October 1951, pages 536–38.

Paul Weiss, "The Prediction Paradox," *Mind*, Vol. 61, April 1952, pages 265–69.

W. V. Quine, "On a So-called Paradox," *Mind*, Vol. 62, January 1953, pages 65–67.

Frank B. Ebersole, "The Definition of 'Pragmatic Paradox,'" *Mind*, Vol. 62, January 1953, pages 80–85.

R. Shaw, "The Paradox of the Unexpected Examination," *Mind*, Vol. 67, July 1958, pages 382–84.

Ardon Lyon, "The Prediction Paradox," *Mind*, Vol. 68, October 1959, pages 510–17.

T. H. O'Beirne, "Can the Unexpected *Never* Happen?," *The New Scientist*, Vol. 15, May 25, 1961, pages 464–65; letters and replies, June 8, 1961, pages 597–98.

David Kaplan and Richard Montague, "A Paradox Regained," *Notre Dame Journal of Formal Logic,* Vol. 1 (1960), pages 79–90.

G. C. Nerlich, "Unexpected Examinations and Unprovable Statements," *Mind,* Vol. 70, October 1961, pages 503–13.

Brian Medlin, "The Unexpected Examination," *American Philosophical Quarterly,* Vol. 1, January 1964, pages 1–7.

B. Meltzer, "The Third Possibility," *Mind,* Vol. 73, July 1964, pages 430–33.

R. A. Sharpe, "The Unexpected Examination," *Mind,* Vol. 74, April 1965, page 255.

B. Meltzer and I. J. Good, "Two Forms of the Prediction Paradox," *British Journal for the Philosophy of Science,* Vol. 16, May 1965, pages 50–51.

J. M. Chapman and R. J. Butler, "On Quine's 'So-called Paradox,' " *Mind,* Vol. 74, July 1965, pages 424–25.

James Kiefer and James Ellison, "The Prediction Paradox Again," *Mind,* Vol. 74, July 1965, pages 426–27.

Judith Schoenberg, "A Note on the Logical Fallacy in the Paradox of the Unexpected Examination," *Mind,* Vol. 75, January 1966, pages 125–27.

J. T. Fraser, "Note Relating to a Paradox of the Temporal Order," *The Voices of Time,* J. T. Fraser, ed. New York: Braziller, 1966, pages 524–26, 679.

J. A. Wright, "The Surprise Exam: Prediction on Last Day Uncertain," *Mind,* Vol. 76, January 1967, pages 115–17.

D. R. Woodall, "The Paradox of the Surprise Examination," *Eureka,* No. 30, October 1967, pages 31–32.

2. KNOTS AND BORROMEAN RINGS

Peter Guthrie Tait, "On Knots," *Scientific Papers* (Cambridge, England: Cambridge University Press, 1898), Vol. 1, pages 273–347.

K. Reidemeister, *Knotentheorie.* New York: Chelsea Publishing, 1948.

R. H. Crowell and R. H. Fox, *Introduction to Knot Theory.* New York: Ginn and Company, 1963.

R. H. Crowell, "Knots and Wheels," *Enrichment Mathematics for High School.* Washington, D.C.: National Council of Teachers of Mathematics, 1963.

M. K. Fort, Jr., ed., *Topology of 3-Manifolds and Related Topics*. New York: Prentice-Hall, 1963.

L. P. Neuwirth, *Knot Groups*. Princeton, N.J.: Princeton University Press, 1965.

3.   THE TRANSCENDENTAL NUMBER e

Constance Reid, *"e," From Zero to Infinity*, 3d edition. New York: Thomas Y. Crowell, 1955. See pages 154-73.

H. V. Baravelle, "The Number e: The Base of Natural Logarithms," *The Mathematics Teacher*, Vol. 38, No. 8, December 1945, pages 350-55.

R. G. Stoneham, "A Study of 60,000 Digits of the Transcendental 'e,'" *American Mathematical Monthly*, Vol. 72, No. 5, May 1965, pages 483-500.

4.   GEOMETRIC DISSECTIONS

Henry Ernest Dudeney, *The Canterbury Puzzles*. New York: Dover Publications (paperback), 1958.

Henry Ernest Dudeney, *Amusements in Mathematics*. New York: Dover Publications (paperback), 1958.

Henry Ernest Dudeney, *536 Puzzles and Curious Problems*, edited by Martin Gardner. New York: Scribner, 1967.

Sam Loyd, *Mathematical Puzzles of Sam Loyd*, 2 vols., edited by Martin Gardner. New York: Dover Publications (paperback), 1959-1960.

V. G. Boltyanskii, *Equivalent and Equidecomposable Figures*, translated and adapted from the first Russian edition (1956) by Alfred K. Henn and Charles E. Watts. Boston: D. C. Heath, Topics in Mathematics Series, 1963.

Harry Lindgren, *Geometric Dissections*. Princeton, N.J.: Van Nostrand, 1964.

Six papers on dissection, written by members of the Mathematics Staff of the College, University of Chicago, in *The Mathematics Teacher*—Vol. 49, May 1956, pages 332-43; October 1956, pages 442-54; December 1956, pages 585-96; Vol. 50, February 1957, pages 125-35; May 1957, pages 330-39; Vol. 51, February 1958, pages 96-104.

Harry Lindgren, "Some Approximate Dissections," *Journal of Recreational Mathematics*, Vol. 1, No. 2, April 1968, pages 79-92.

5. SCARNE ON GAMBLING

John Scarne and Clayton Rawson, *Scarne on Dice.* Harrisburg, Pa.: Military Service Publishing Co., 1945.

John Scarne, *The Amazing World of John Scarne.* New York: Crown Publishers, 1956.

John Scarne, *The Odds Against Me.* New York: Simon and Schuster, 1966.

Oswald Jacoby, *How to Figure the Odds.* Garden City, N.Y.: Doubleday, 1947.

Oswald Jacoby, *Oswald Jacoby on Gambling.* Garden City, N.Y.: Doubleday, 1963.

Roger Baldwin, *Playing Blackjack to Win.* New York: M. Barrows, 1957.

Sidney H. Radner, *Roulette and Other Casino Games.* New York: Wehman Brothers, 1958.

Edward O. Thorp, *Beat the Dealer.* New York: Blaisdell, 1962.

Frank Garcia, *Marked Cards and Loaded Dice.* New York: Prentice-Hall, 1962.

Alan H. Wilson, *The Casino Gambler's Guide.* New York: Harper & Row, 1965.

6. THE CHURCH OF THE FOURTH DIMENSION

Charles Howard Hinton, *The Fourth Dimension.* London: Swan Sonnenschein, 1904; Allen & Unwin, 1951.

E. H. Neville, *The Fourth Dimension.* Cambridge, England: Cambridge University Press, 1921.

Henry Parker Manning, *The Fourth Dimension Simply Explained.* New York: Munn, 1910; New York: Dover Publications (paperback), 1960.

Henry Parker Manning, *Geometry of Four Dimensions.* New York: Macmillan, 1914; Dover Publications (paperback), 1956.

D. M. Y. Sommerville, *An Introduction to the Geometry of N Dimensions.* London: Methuen, 1929; New York: Dover Publications (paperback), 1958.

Karl Heim, *Christian Faith and Natural Science.* New York: Harper, 1953; Harper (paperback), 1957.

8. A MATCHBOX GAME-LEARNING MACHINE

W. Ross Ashby, "Can a Mechanical Chess-player Outplay Its Designer?" *British Journal for the Philosophy of Science,* Vol. 3 (1952), No. 44.

Claude E. Shannon, "Game Playing Machines," *Journal of the Franklin Institute,* Vol. 260, No. 6, December 1955, pages 447–53.

Allen Newell, J. C. Shaw, and H. A. Simon, "Chess-playing Programs and the Problem of Complexity, *IBM Journal of Research and Development,* Vol. 2, No. 4, October 1958, pages 320–35.

Alex Bernstein and Michael de V. Roberts, "Computer v. Chess-player," *Scientific American,* June 1958, pages 96–105.

A. L. Samuel, "Some Studies in Machine Learning, Using the Game of Checkers," *IBM Journal of Research and Development,* Vol. 3, No. 3, July 1959, pages 210–29.

John Maynard Smith and Donald Michie, "Machines That Play Games," *The New Scientist,* No. 260, November 9, 1961, pages 367–69.

Allen Newell and H. A. Simon, "Computer Simulation of Human Thinking," *Science,* Vol. 134 (1961), pages 2011–17.

N. Nilsson, *Learning Machines.* New York: McGraw-Hill, 1965.

9. SPIRALS

E. H. Lockwood, *A Book of Curves.* Cambridge, England: Cambridge University Press, 1961.

William Kingdon Clifford, *The Common Sense of the Exact Sciences.* New York: Dover Publications, 1955. See pages 152–64.

Theodore Andrea Cook, *Spirals in Nature and Art.* London: J. Murray, 1903.

Martin Gardner, *The Ambidextrous Universe,* New York: Basic Books, 1964; London: Allen Lane, 1967.

10. ROTATIONS AND REFLECTIONS

Charles Carroll Bombaugh, *Oddities and Curiosities of*

Words and Literature, edited by Martin Gardner. New York: Dover Publications, 1961. See pages 345–46.

Peter Newell, *Topsys & Turvys.* New York: Dover Publications, 1964.

The Incredible Upside-Downs of Gustave Verbeek, edited by George M. Naimark. Summit, N.J.: The Rajah Press, 1963.

11. PEG SOLITAIRE

References dealing primarily with problems:

J. Busschopp, *Recherches sur le jeu du solitaire.* Bruges, 1879.

Berkeley (pseudonym of W. H. Peel), *Dominoes and Solitaire.* London: G. Bell, 1890; New York: Frederick A. Stokes, 1890.

Ernest Bergholt, *The Game of Solitaire.* London: George Routledge, 1920; New York: Dutton, 1921.

Paul Redon, *Le Jeu de solitaire.* Paris, no date.

W. Ahrens, *Mathematische Unterhaltungen und Spiele.* Berlin: Druck und Verlag von B. G. Teubner, 1910. See Vol. 1, Chapter 8.

T. R. Dawson, "Solitaire," *Fairy Chess Review,* Vol. 5, June 1943, pages 42–43. Later issues contain other solitaire problems by Dawson, T. H. Willcocks and others.

Harry O. Davis, "33-Solitaire: New Limits, Small and Large," *Mathematical Gazette,* Vol. 51, May 1967, pages 91-100.

Donald C. Cross, "Square Solitare and Variations," *Journal of Recreational Mathematics,* Vol. 1, No. 2, April 1968, pages 121–123.

References dealing primarily with theory:

M. Reiss, "Beiträge zur Theorie des Solitär-Spiels," *Crelles Journal* (Berlin), Vol. 54 (1857), pages 344–79.

Gaston Tissandier, *Popular Scientific Recreations* (translation of an 1881 French work). London: Ward, Lock and Bowden, 1882. See pages 735–39.

A. M. H. Hermary, "Le Jeu du solitaire," *Récréations Math-*

ématiques, edited by Édouard Lucas. Paris: Blanchard (paperback), 1960. Reprint of 1882 edition. See Vol. 1, pages 87–141.

G. Kowalewski, "Das Solitärspiel," *Alte und neue mathematische Spiele*. Leipzig: Teubner, 1930.

B. M. Stewart, "Solitaire on a Checkerboard," *American Mathematical Monthly*, April 1941, pages 228–33.

B. M. Stewart, *Theory of Numbers*, revised edition. New York: Macmillan 1964. See Chapter 2.

Mannis Charosh, "Peg Solitaire," *The Mathematics Student Journal*, Vol. 9, March 1962, pages 1–3.

J. D. Beasley, "Some Notes on Solitaire," *Eureka*, No. 25, October 1962, pages 13–28.

12. FLATLANDS

Charles Howard Hinton, "A Plane World," *Scientific Romances* (London: Allen & Unwin, 1888), Vol. I, pages 135–59.

Charles Howard Hinton, *A New Era of Thought*. London: Swan Sonnenschein, 1888.

Charles Howard Hinton, *The Fourth Dimension*. London: Swan Sonnenschein, 1904; Allen & Unwin, 1934.

Charles Howard Hinton, *An Episode of Flatland*. London: Swan Sonnenschein, 1907.

Edwin Abbott Abbott, *Flatland*, with Introduction by Banesh Hoffmann, New York: Dover Publications, 1952.

Dionys Burger, *Sphereland*. New York: Crowell, 1965. A translation of a Dutch physicist's sequel to Abbott's *Flatland*. The story is told by A. Hexagon, grandson of Abbott's A. Square.

13. CHICAGO MAGIC CONVENTION

Royal V. Heath, *Mathemagic*. New York: Simon and Schuster, 1933; Dover Publications (paperback), 1953.

Wallace Lee, *Math Miracles*. Privately printed, 1950; revised edition, 1960.

Martin Gardner, *Mathematics, Magic and Mystery*. New York: Dover Publications, 1956.

William Simon, *Mathematical Magic*. New York: Scribner, 1964.

14. TESTS OF DIVISIBILITY

Leonard Eugene Dickson, "Criteria of Divisibility by a Given Number," *History of the Theory of Numbers* (New York: Chelsea Publishing, 1952), Vol. I, Chapter 12, pages 337–46. Originally published in 1919.

J. M. Elkin, "A General Rule for Divisibility Based on the Decimal Expansion of the Reciprocal of the Divisor," *American Mathematical Monthly*, Vol. 59, May 1952, pages 316–18.

Kenneth A. Seymour, "A General Test for Divisibility," *The Mathematics Teacher*, Vol. 56, March 1963, pages 151–54.

J. Bronowski, "Division by 7," *Mathematical Gazette*, Vol. 47, October 1963, pages 234–35.

Wendell M. Williams, "A Complete Set of Elementary Rules for Testing for Divisibility," *The Mathematics Teacher*, Vol. 56, October 1963, pages 437–42.

Jack M. Elkin, "Repeating Decimals and Tests for Divisibility," *The Mathematics Teacher*, Vol. 57, April 1964, pages 215–18.

E. A. Maxwell, "Division by 7 or 13," *Mathematical Gazette*, Vol. 49, February 1965, page 84.

Sister M. Barbara Stastny, "Divisibility Patterns in Number Bases," *The Mathematics Teacher*, Vol. 58, April 1965, pages 308–10.

Benjamin Bold, "A General Test for Divisibility by Any Prime (Except 2 and 5)," *The Mathematics Teacher*, Vol. 58, April 1965, pages 311–12.

Henri Feiner, "Divisibility Test for 7," *The Mathematics Teacher*, Vol. 58, May 1965, pages 429–32.

John Q. Jordan, "Divisibility Tests of the Noncongruence Type," *The Mathematics Teacher*, Vol. 58, December 1965, pages 709–12.

Robert Pruitt, "A General Divisibility Test," *The Mathematics Teacher*, Vol. 59, January 1966, pages 31–33.

R. L. Morton, "Divisibility by 7, 11, 13, and Greater Primes," *The Mathematics Teacher*, Vol. 61, April 1968, pp. 370–73.

16. THE EIGHT QUEENS AND OTHER CHESSBOARD DIVERSIONS

The problem of the nonattacking queens:

A. Ahrens, *Mathematische Unterhaltungen und Spiele*. Leipzig: Teubner, 1910. Vol. I, Chapter 9.

W. W. Rouse Ball, *Mathematical Recreations and Essays*, revised edition. New York: Macmillan, 1960. Chapter 6.

Maurice Kraitchik, *Mathematical Recreations*, revised edition. New York: Dover Publications, 1953. Chapter 10.

Édouard Lucas, ed., *Récréations Mathématiques*. Paris: Blanchard, 1960. Chapter 4. A reprint of the original 1882 edition.

The problem of the nonattacking rooks:

Henry Ernest Dudeney, *Amusements in Mathematics*. London: Thomas Nelson and Sons, 1917. New York: Dover Publications, 1958. Pages 76, 88 (problem 296), and 96.

A. M. Yaglom and I. M. Yaglom, *Challenging Mathematical Problems with Elementary Solutions*. San Francisco: Holden-Day, 1964. Section III.

Joseph S. Madachy, *Mathematics on Vacation*. New York: Scribner, 1966. Chapter 2.

17. A LOOP OF STRING

Caroline Furness Jayne, *String Figures*. New York: Scribner, 1906; Dover Publications (paperback), 1962.

W. W. Rouse Ball, *An Introduction to String Figures*. Cambridge, England: Cambridge University Press, 1920. Reprinted in *String Figures and Other Monographs* (New York: Chelsea Publishing, 1960.)

Kathleen Haddon, *Artists in String*. London: Methuen, 1930; New York: Dutton, 1930.

Kathleen Haddon, *String Games for Beginners*. Cambridge, England: Cambridge University Press, 1934.

Joseph Leeming, *Fun with String*. New York: Stokes, 1940.

Jerome Barry, *Leopard Cat's Cradle*. Garden City, N.Y.: Doubleday, 1942.

Eric Franklin, *Kamut: Pictures in String*. New York: Arcas, 1945.

18. CURVES OF CONSTANT WIDTH

Franz Reuleaux, *The Kinematics of Machinery*. New York: Macmillan, 1876; Dover Publications, 1964. See pages 129–46.

Wilhelm Blaschke, *Kreis und Kugel*. Leipzig, 1916; Berlin: W. de Gruyter, 1956.

Hans Rademacher and Otto Toeplitz, *The Enjoyment of Mathematics*. Princeton, N.J.: Princeton University Press, 1957. See pages 163–77, 203.

I. M. Yaglom and V. G. Boltyanskii, *Convex Figures*. New York: Holt, Rinehart & Winston, 1961. Chapters 7 and 8.

J. H. Cadwell, *Topics in Recreational Mathematics*. Cambridge, England: Cambridge University Press, 1966. Chapter 15.

Michael Goldberg, "Trammel Rotors in Regular Polygons," *American Mathematical Monthly*, Vol. 64, February 1957, pages 71–78.

Michael Goldberg, "Rotors in Polygons and Polyhedra," *Mathematical Tables and Other Aids to Computation*, Vol. 14, July 1960, pages 229–39.

Michael Goldberg, "N-Gon Rotors Making N + 1 Contacts with Fixed Simple Curves," *American Mathematical Monthly*, Vol. 69, June-July 1962, pages 486–91.

On Kakeya's needle problem:

I. M. Yaglom and V. G. Boltyanskii, *Convex Figures*. New York: Holt, Rinehart & Winston, 1961. See pages 61–62, 226–27.

J. H. Cadwell, *Topics in Recreational Mathematics*. Cambridge, England: Cambridge University Press, 1966. See pages 96–99.

A. S. Besicovitch, "The Kakeya Problem," *American Mathematical Monthly*, Vol. 70, August-September 1963, pages 697–706.

A. A. Blank, "A Remark on the Kakeya Problem," *American*

Mathematical Monthly, Vol. 70, August-September 1963, pages 706–11.

19. REP-TILES: REPLICATING FIGURES ON THE PLANE

C. Dudley Langford, "Uses of a Geometrical Puzzle," *Mathematical Gazette,* Vol. 24, July 1940, pages 209–11.

R. Sibson, "Comments on Note 1464," *Mathematical Gazette,* Vol. 24, December 1940, page 343.

Howard D. Grossman, "Fun with Lattice Points," *Scripta Mathematica,* Vol. 14, June 1948, pages 157–59.

Solomon W. Golomb, "Replicating Figures on the Plane," *Mathematical Gazette,* Vol. 48, December 1964, pages 403–12.

M. Goldberg and B. M. Stewart, "A Dissection Problem for Sets of Polygons," *American Mathematical Monthly,* Vol. 71, December 1964, pages 1077–95.

Roy O. Davies, "Replicating Boots," *Mathematical Gazette,* Vol. 50, May 1966, page 157.